The Minerals, Metals & Materials Series

Brian Wisner · Simona E. Hunyadi Murph ·
Ioannis N. Mastorakos · Muralidharan Paramsothy
Editors

Composite Materials

Sustainable and Eco-Friendly Materials and Application

TMS

Springer

Editors
Brian Wisner
Ohio University
Athens, OH, USA

Simona E. Hunyadi Murph
Savannah River National Laboratory
Aiken, SC, USA

Ioannis N. Mastorakos
Clarkson University
Potsdam, NY, USA

Muralidharan Paramsothy
NanoWorld Innovations
Singapore, Singapore

ISSN 2367-1181 ISSN 2367-1696 (electronic)
The Minerals, Metals & Materials Series
ISBN 978-3-031-50179-1 ISBN 978-3-031-50180-7 (eBook)
https://doi.org/10.1007/978-3-031-50180-7

This Springer imprint is published by the registered company Springer Nature Switzerland AG
The registered company address is: Gewerbestrasse 11, 6330 Cham, Switzerland

Paper in this product is recyclable.

Contents

About the Editors

Brian Wisner received his bachelor's degrees in Mechanical Engineering and Physics from Widener University before moving to Drexel University to complete his master's and Ph.D. in Mechanical Engineering and Mechanics. During his work there he studied the mechanical response of metals and composite materials used in the aerospace industry, attempting to understand the role that microstructure plays in global properties. In 2019, he joined Ohio University as an Assistant Professor of Mechanical Engineering with a focus in Manufacturing and Nondestructive Evaluation. He works on projects including understanding the microscale behavior of coal waste used as a filler for polymer composites, the manufacturing of cost-effective batteries using materials easier to obtain than lithium ion. He also works to improve sustainability of legacy components by leveraging nondestructive testing to track damage and predict component life. Other projects include sustainable additive manufacturing methods to create needed components safely and sustainably.

Dr. Wisner has been an active member of TMS since 2015 and served on the Composite Materials Committee since 2020 as Vice Chair and now Chair and has organized the Eco-Friendly Composite Symposium for three years. Additionally, he has organized a special issue of *JOM* on this topic.

Simona E. Hunyadi Murph is a Senior Fellow at Savannah River National Laboratory (SRNL) and the Technical Liaison for the Department of Energy-Office of Environmental Management (DOE-EM), MSIP Program. She concurrently holds an Adjunct Professor appointment at the University of Georgia. In recent years, she has served on an intergovernmental personnel assignment with the DOE-EM and has managed SRNL's Laboratory Directed Research and Development program.

Dr. Murph's nanoscience and nanotechnology research efforts have produced cutting-edge scientific and technological innovations to benefit the nation's priorities and the U.S. Department of Energy's initiatives. She has authored an impressive body of literature with over 170 publications and more than 6,300 citations. Apart from the extensive list of publications and citations, a key indicator of the impact of her work is 20 issued or pending patents. Dr. Murph is the recipient of many prestigious awards including the 2023 SRNL Battelle Inventor of the Year for significant scientific or engineering contributions that have important societal and financial impact. She is also the recipient of the 2023 Brimacombe Medal from The Minerals, Metals and Materials Society (TMS) for her cutting-edge scientific and engineering innovation of anisotropic and shape-selective nanomaterials, extraordinary effort in developing the next generation of leaders, and dedicated service to TMS. She has received noteworthy awards in the past including Principal Investigator of the Year (NNSA), U.S. Clean Energy Education and Empowerment (C3E) Award Finalist, Inspirational Woman in STEM recognition (DOE), Distinguished Alumna and Presidential Alumna Award, and several SRNL Director's Awards for Exceptional Achievements among others.

Dr. Murph holds a Ph.D. in Chemistry from the University of South Carolina, an Education Specialist (Ed.S.) in Educational Leadership from Augusta University, and both an M.S. in Chemistry and B.S. in Chemistry/Physics with a minor in Education from Babes-Bolyai University, Romania.

Ioannis N. Mastorakos specializes in materials science and mechanical engineering. He holds a Ph.D. in engineering from Aristotle University of Thessaloniki, Greece, and has a background in physics. He is an Associate Professor in the Department of Mechanical and Aerospace Engineering at Clarkson University. His participation in various faculty committees, including undergraduate recruitment, faculty search, and marketing, emphasizes his engagement in shaping the educational experience of future engineers. Dr. Mastorakos' research focuses on the multiscale modeling of materials, especially in composite metallic nanostructures and high entropy alloys. His work was funded by the US-DOE, NSF, NASA, and the NYS Center for Advanced Materials Processing. He has contributed to various peer-reviewed journals, conferences, and seminars, has organized various symposia, and is recognized for his work on high-strength metal nanocomposites. He is involved in various professional organizations such as the American Society of Mechanical Engineering, Materials Research Society, and The Minerals, Metals & Materials Society, where he is currently the vice-chair of the Composite Materials Committee.

Muralidharan Paramsothy is an early-retired scientist based in Singapore. Before retiring by choice, he was active for almost 5 years in supporting condition monitoring as well as intelligent energy projects in the Agency for Science, Technology and Research (A*STAR). Prior to that, he was active for almost 10 years in contributing to nanomaterials and nano(science)technology endeavors in Singapore. Overall, he has extensively managed academic as well as industrial stakeholders. He has numerous internationally published peer-reviewed articles, the Nanoscale Electro Negative Interface Density or NENID theory, and intellectual property pertaining to optical fiber sensor utilization to his name. He has organized a fair number of academic symposia in the United States and Canada. He has also served as a guest editor for themed topics in journals. He is driven by work that benefits one and all, not work that promotes only the individual.

Brian Wisner · Simona E. Hunyadi Murph ·
Ioannis N. Mastorakos · Muralidharan Paramsothy
Editors

Composite Materials

Sustainable and Eco-Friendly Materials and Application

TMS

Springer

Analysis of the Incorporation of Industrial Granite Waste in an Epoxy Matrix on Compression Performance

D. C. R. Velasco, J. A. T. Linhares Júnior, F. P. Lopes, M. T. Marvila, A. G. de Azevedo, and C. M. Vieria

Abstract The increasing concern about waste generation in the industry also encompasses the granite cutting sector. In this context, there is an interest in exploring the potential of these waste materials for applications in new materials (composites). This study aims to evaluate how the incorporation of granite particulates affects the compressive properties of an epoxy system, as well as to characterize the utilized waste material. The waste material was characterized using laser granulometry. Subsequently, the particulates were used in the fabrication of polymers composites at different volumetric fractions of waste (0, 12.5, 25, 37.5, and 50%). These composites were tested according to ASTM D695 and subsequently analyzed through Analysis of Variance and Tukey's test. The obtained results provided information regarding the particle size distribution of the waste material, as well as the impact of these wastes on strength of the polymer matrix composites.

Keywords Composites · Mechanical properties · Epoxy

D. C. R. Velasco (✉) · J. A. T. L. Júnior · F. P. Lopes · C. M. Vieria
UENF—State University of the Northern Rio de Janeiro, LAMAV—Advanced Materials Laboratory, Av. Alberto Lamego, 2000, Campos dos Goytacazes, Rio De Janeiro 28013-602, Brazil
e-mail: davidc.r.v2014@gmail.com

C. M. Vieria
e-mail: markssuel.marvila@ufv.br

A. G. de Azevedo
UENF—State University of the Northern Rio de Janeiro, LECIV—Civil Engineering Laboratory, Av. Alberto Lamego, 2000, Campos dos Goytacazes, Rio De Janeiro 28013-602, Brazil
e-mail: afonso@uenf.br

M. T. Marvila
UFV—Federal University of Viçosa Campus Rio Paranaíba (UFV-CRP), Rodovia BR 230 KM 7, Rio Paranaíba 38810-000, Brazil
e-mail: markssuel.marvila@ufv.br

B. Wisner et al. (eds.), *Composite Materials*, The Minerals, Metals & Materials Series,
https://doi.org/10.1007/978-3-031-50180-7_1

Introduction

The concept of sustainable development has existed for many years, but it was only after the Second World War that society began to give due importance to it, since with technology there was the possibility of developing and meeting human needs in a safe way, without harming the environment. As society develops and the population grows, needs tend to increase in equal proportion and within the productive context, a lot of waste is being generated, with the need to seek the best forms of management, so that the next generations do not have severe problems [1].

Brazil is an important producer of dimension stones on the international scene. In the productive context, a large amount of waste is generated in the various processing stages, as shown in Fig. 1. These wastes can pollute the air and contaminate the soil, causing damage to the environment [2, 3].

Granite is the main rock produced in Brazil and is widely used in civil construction and in its processing large amounts of waste are generated, especially in the polishing steps. These residues tend to accumulate over time, occupying open-air patios. In the current context, around 25% of the total area of the industrial installations that process granite are occupied with this material that needs an adequate destination. Granite residue can be applied in composites, with the potential to improve mechanical properties and reduce production costs [4–6].

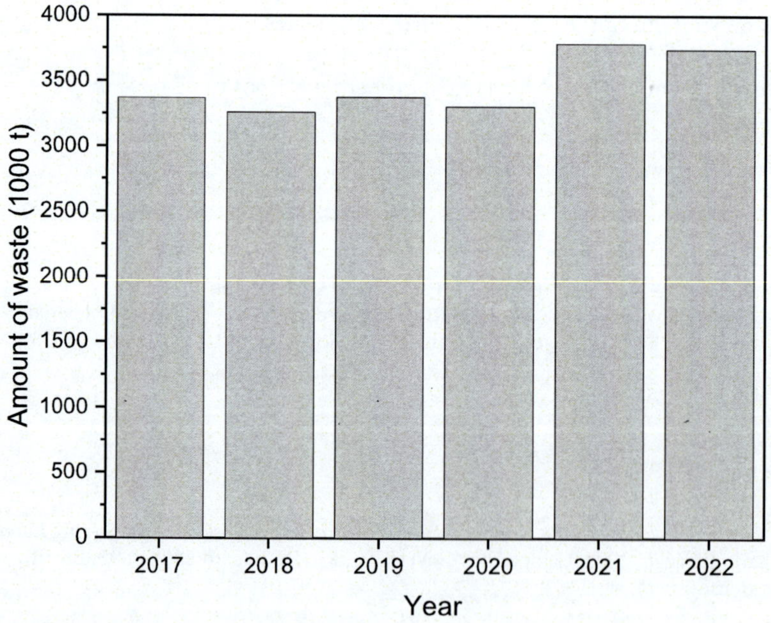

Fig. 1 Amount of waste from the processing of ornamental stones [4]

Polymeric matrix composite materials have several contrasts in relation to better known groups such as polymers, metals, and ceramics. These materials have, in general, high resistance, rigidity, and still have a low weight. These characteristics combined are of great importance for engineering design [7–9].

The study by Dubey et al. [10] investigated the behavior of epoxy matrix composites with the addition of granite particulate, analyzing the impact resistance (Izod), Vickers hardness, and water absorption. The best performances were achieved with 30 to 40% residue. Gonçalves et al. [11] developed epoxy matrix composites with 0% (reference), 30%, and 50% mass addition of particulates from granite waste, the incorporation of 50% achieved the best resistance results, reaching 79 MPa.

In terms of using granite residue in an epoxy resin matrix, the objective of this work was to characterize the granulometry of the granite residue, as well as to elaborate epoxy matrix composites in different volumetric percentages of particulate matter: 0, 12.5, 25, 37.5, and 50%. Compressive strength was studied for each fraction. Analysis of variance (ANOVA) and the Tukey test were used to statistically verify whether there was a significant difference between the sample groups.

Materials and Methods

The materials used in carrying out this study are epoxy resin as a matrix and granite particulate. The epoxy system used was SQ 1005/SQ 3131, composed of bisphenol A diglycidyl ether (DGEBA) with diethylene triamine hardener (DETA). The granite residue was sieved at 100 Mesh. In the next step, it was characterized in terms of its granulometry, using a laser granulometer, Microtrac S2500.

The compositions were prepared with a stoichiometric ratio equal to 20 phr. The particulate was incorporated into the resin at 0, 12.5, 25, 37.5, and 50% (vol.). To make the specimens, silicone molds measuring 12.7 × 12.7 × 6 mm were filled with the proposed mixtures.

The specimens were submitted to compression tests, following the procedure described by the ASTM D695-15 [12] standard, at a speed of 1mm/min. The equipment used to carry out the tests was an INSTRON universal mechanical testing machine, model 34FM-100-AS, Fig. 2.

After carrying out the compression tests, the data were statistically evaluated using analysis of variance (ANOVA) and the Turkey test, in order to verify whether there was a statistically significant difference between the groups of samples. A significance level of 5% was adopted and the analyses were performed in the PAST software, following the Copenhagen and Holland algorithm [13].

Fig. 2 Testing machine, INSTRON, model 34FM-100-AS

Results and Discussion

Figure 3 shows the granite particulate size distribution curve. The analysis reveals that the average size is approximately 29 μm, which is notably smaller than the sieve opening, which is 150 μm. Besides, it is important to highlight that a probability of finding particles larger than the sieve opening was detected, suggesting that at least a portion of the particles do not have a spherical shape.

The results of this work made it possible to obtain the compressive strength and modulus of elasticity, which can be seen in Table 1.

Through these results, it was possible to generate Fig. 4, as well as to perform ANOVA and Tukey's test. Through these, the insertion of granite particles initially did not lead to significant variations in compressive strength. However, in the formulation

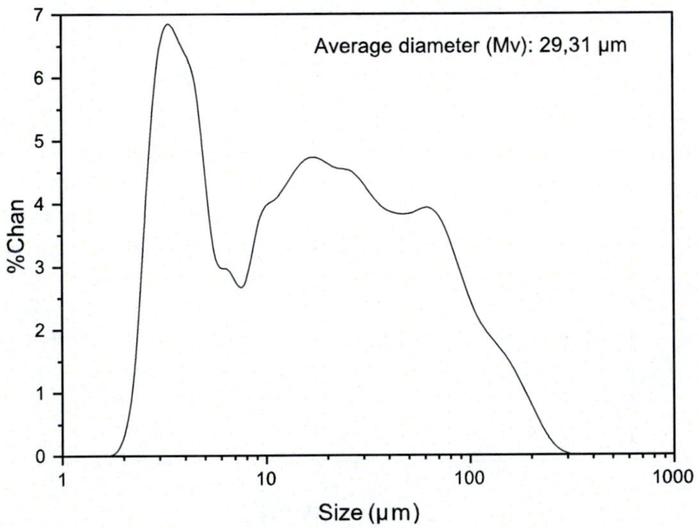

Fig. 3 Compressive strength as a function of the amount of residue

Table 1 Results obtained in the compression test

Granite (%)	Compressive strength (MPa)	Modulus of elasticity (MPa)
0	82.09 ± 5.72	2018.4 ± 313.4
12.5	77.33 ± 4.7	1753.8 ± 171
25	76.18 ± 4.96	1670.2 ± 150.9
37.5	77.1 ± 5.55	1891 ± 152.5
50	68.18 ± 4.57	1802 ± 330.4

with 50% particulates, a change in behavior was observed, with a significant loss of performance. This can be attributed to particulates being agglomerated and having a poorer interface with the matrix [14]. It should also be added that the number of particulates favors an increase in viscosity, which can harm the mixture with the hardener and generate regions with poor intercrosslinks [15].

Regarding the modulus of elasticity, Fig. 5, no significant variations in stiffness were identified according to the analysis of variance (ANOVA). This observation can be attributed to the fact that the performance of particles is not only related to their nature, but also to their size and shape. It is possible to obtain both gains and reductions in stiffness when incorporating residuals, as observed in a previous study [11].

However, even if the incorporation of waste has not resulted in gains in strength or variations in stiffness, this proves to be beneficial both in terms of environmental impact and potential cost reductions, which is the objective of several studies in the literature [16, 17].

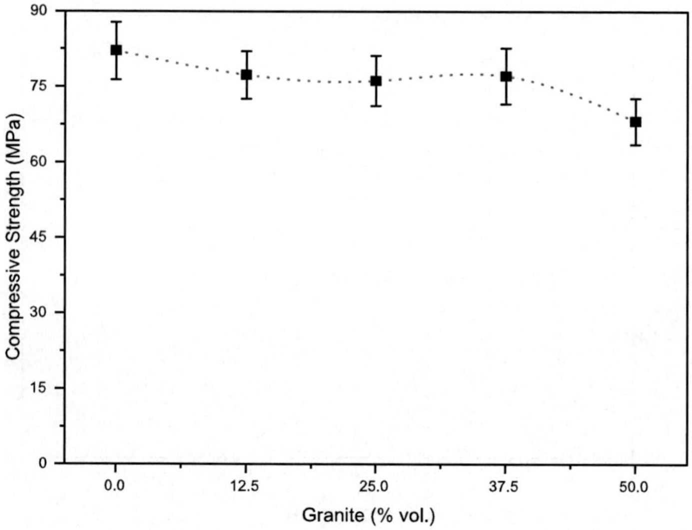

Fig. 4 Compressive strength as a function of the amount of waste

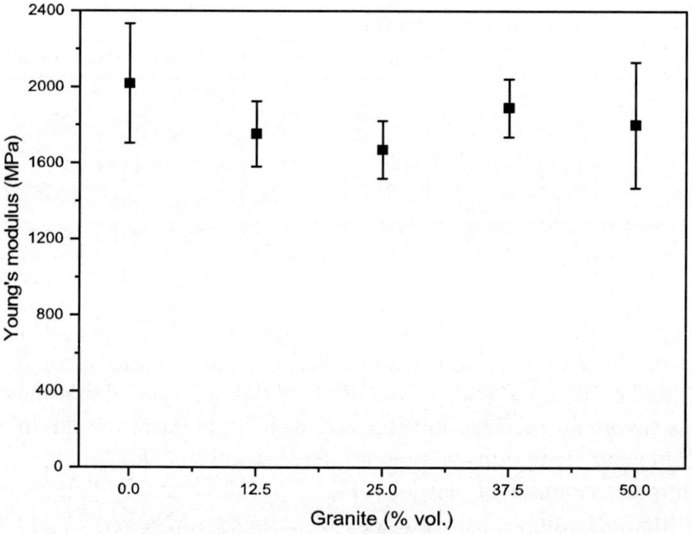

Fig. 5 Young modulus as a function of the amount of waste

Conclusions

The results of this study allowed a detailed characterization of the granite waste and revealed the influence of its incorporation in composites with a polymeric matrix. It was notable the finding that it is feasible to incorporate up to 37.5% of waste without observing significant variations in mechanical properties, which not only opens doors to environmental gains, but also to potential cost savings, making the approach ecologically conscious and economically advantageous.

References

1. Dubey SC, Mishra V, Sharma A (2021) A review on polymer composite with waste material as reinforcement. Mater Today: Proc, 2846–2851. https://doi.org/10.1016/j.matpr.2021.03.611
2. Silva TLC, Carvalho EAS, Barreto GNS, Silva TBP, Demartini TJC, Vieira CMF (2023) Characterization of artificial stone developed with granite waste and glass waste in epoxy matrix. J Mater Res Technol 26:2528–2538. https://doi.org/10.1016/j.jmrt.2023.08.045
3. Nayak SK, Satapathy A, Mantry S (2022) Use of waste marble and granite dust in structural applications: a review. J Build Eng 46:103742. https://doi.org/10.1016/j.jobe.2021.103742
4. Associação Brasileira da Indústria de Rochas Ornamentais (2022) Balanço das exportações e importações de rochas ornamentais em 2022. http://comexstat.mdic.gov.br
5. Shamsabadi EA, Ghalehnovi M, Brito J, Khodabakhshian A (2018) Performance of concrete with waste granite powder: the effect of superplasticizers. Appl Sci (Switzerland) 8. https://doi.org/10.3390/app8101808
6. Reddy BM, Reddy RM, Reddy BCM, Reddy PV, Rao HR, Reddy YVM (2020) The effect of granite powder on mechanical, structural and water absorption characteristics of alkali treated cordia dichotoma fiber reinforced polyester composite. Polym Test 91:106782. https://doi.org/10.1016/j.polymertesting.2020.106782
7. Maluga R, Kumar MS, Pati R, Kumar SS (2023) Physical, mechanical and wear characterization of epoxy composites reinforced with granite/marble powder. Mater Today Proc. https://doi.org/10.1016/j.matpr.2023.02.045
8. Daniyan IA, Mpofu K, Adeodu AO, Adesina O (2021) Development of carbon fibre reinforced polymer matrix composites and optimization of the process parameters for railcar applications. Mater Today: Proc 38(2):628–634. https://doi.org/10.1016/j.matpr.2020.03.480
9. Rashid MA, Zhu S, Zhang L, Jin K, Liu W (2023) High-performance and fully recyclable epoxy resins cured by imine-containing hardeners derived from vanillin and syringaldehyde. Euro Polym J 187:111878. https://doi.org/10.1016/j.eurpolymj.2023.111878
10. Subhash C, Krishna R, Raj S, Sai H, Rao R (2018) Development of granite powder reinforced epoxy composites. Mater Today: Proc 5(5):2214–7853. https://doi.org/10.1016/j.matpr.2018.02.286
11. Gonçalves JAV, Campos DAT, Oliveira GDJ, Rosa MDLS, Macêdo MA (2014) Mechanical properties of epoxy resin based on granite stone powder from the sergipe fold-and-thrust belt composites. Mater Res 17:878–887. https://doi.org/10.1590/S1516-14392014005000100
12. American Society for Testing and Materials (2023) D695: standard test method for compressive properties of rigid plastics, West Conshohocken. https://doi.org/10.1520/D0695-23
13. Holland B (1988) Computation of the distribution of the maximum studentized range statistic with application to multiple significance testing of simple effects. J Stat Comput Simul 30:1–15. https://doi.org/10.1080/00949658808811082
14. Luo X, Gao J, Liu C, He Z (2021) Effects of polymers on workability and early microstructure of gypsum-based materials. ACI Mater J. https://doi.org/10.14359/51732929

15. Velasco DCR, Lopes FPD, Souza D, Lopera HAC, Monteiro SN, Vieira CMF (2023) Evaluation of composites reinforced by processed and unprocessed coconut husk powder. Polymers (Basel) 15:1195. https://doi.org/10.3390/polym15051195
16. Velasco DCR, Lopes FPD, Simonassi NT, Vieira CMF (2022) Influência da incorporação de resíduos na permeabilidade ao vapor de água de compósitos de matriz epoxídica/Influence of the incorporation of waste on the water vapor permeability of epoxide matrix composites. Brazilian J Develop 8:24272–24282. https://doi.org/10.34117/bjdv8n4-101
17. Brito MR, Marvila MT, Linhares JAT, Azevedo ARG (2023) Evaluation of the properties of adobe blocks with clay and manure. Buildings 13:657. https://doi.org/10.3390/buildings13030657

Design and Characterization of Kevlar/Epoxy Composites Infused with Silicon Carbide (SiC) – Zinc Oxide (ZnO) Nanofillers

K. Bupesh Kumar and B. M. Rajaprakash

Abstract Fibre reinforced polymer (FRP) composites are prominent materials used for manufacturing various products due to their distinct properties. It has been observed that the addition of nanofillers improved various properties of the FRP composites. Very few researchers have combined more than one nanofiller for producing FRP composites. The objective of the present work is to study the feasibility of combined effect of silicon carbide (SiC)–zinc oxide (ZnO) nanofillers in the development of Kevlar/Epoxy composite and examine their effect on the various mechanical properties such as tensile strength, flexural strength, and inter laminar shear strength. The composite material infused with the above-mentioned nanofillers was fabricated by using three different weight percentages of both the nanofillers. The experimental results showed that the addition of the SiC-ZnO nanofillers with the weight percentages of 0.75% and 1%, respectively, to the Kevlar/Epoxy composite exhibited higher tensile, flexural, and interlaminar shear strength.

Keywords Fibre reinforced polymer · Nanofillers · Vacuum bag molding · Mechanical characterization

Introduction

Polymer matrix composites, also known as fibre reinforced plastics, are composite materials whose reinforcement phase is made of fibres and their matrix phase is made of polymers. Fibre reinforced polymer (FRP) composites are used in manufacturing

K. Bupesh Kumar (✉)
Department of Mechanical Engineering, University Visvesvaraya College of Engineering, Bangalore University, Bangalore, Karnataka 560 001, India
e-mail: bupesh.bk@gmail.com

B. M. Rajaprakash
Department of Mechanical Engineering, University of Visvesvaraya College of Engineering (UVCE), Bangalore, Karnataka 560 001, India
e-mail: rajaprakashbm@uvce.ac.in

B. Wisner et al. (eds.), *Composite Materials*, The Minerals, Metals & Materials Series,
https://doi.org/10.1007/978-3-031-50180-7_2

various products due to their distinct advantages, namely, lower weight, higher stiffness, higher strength, ability to mold into complex shapes, better corrosion resistance, and damping properties [1]. In recent years, it has been found that the properties of the composites can be altered by altering its components in the nanometer range itself; hence, polymer composites filled with nanofillers which are often referred as nanocomposites are increasingly used in several industrial applications. It has been found that the addition of nanofillers increase the mechanical properties of the material to a significant extent [2]. The polymer composite market is mainly dominated by the aerospace, automobile, defense, aircraft, and marine application. Other applications include oil and gas pipelines, thin film capacitors, fuel tanks, impeller blades, and gas barriers [3].

Researchers in earlier studies have conducted investigations on the role of various nanofillers on the mechanical properties of various fibre reinforced polymer composites. Sujesh. G and C. Ganesan studied the effect of inclusion of nano-silica particle in the glass fibre reinforced plastic manufactured by vacuum bagging technique and found that the addition of nanoparticle increased the mechanical properties such as tensile strength, tensile modulus, and ultimate tensile load [4]. Kumar et al., studied the effect of addition of nano-TiO_2 filler to glass fibre reinforced polymer and concluded that it leads to an increase in the flexural strength and flexural modulus [5]. Raghavendra et al., studied the influence of nano-alumina (Al_2O_3) on the mechanical properties of glass/epoxy laminates and found that the composite with the nanofiller showed better flexural and tensile results when compared with neat composite [6]. Panse et al., investigated the effect of nano-silica in glass fibre and carbon fibre reinforced polymer and found that addition of nano-silica increased the compressive strength of composites [7]. Rajmohan et al., studied the effect of nano-CuO particles on the mechanical properties of glass fibre reinforced plastics and concluded that with increase in the percentage of nanofillers, tensile strength and flexural strength also increased [8].

Researchers also found that the addition of nanofillers up to only a certain weight percentage leads to an enhancement in the mechanical properties of the fibre reinforced composites. Raju et al., studied the mechanical properties of glass fibre reinforced polyester with the presence of 1, 2, 4, and 6% weight of ZnO nanoparticles and concluded that the addition of ZnO nanofillers up to 2% increased the tensile strength and the hardness of the composites and further addition causes decrease in both hardness and tensile strength [9]. Bose et al., investigated the effect of ZnO nanoreinforcements of different weight percentages such as neat, 1, 3, and 5% in epoxy matrix on glass fibre reinforced polymer composite. It was found that the 1% loaded ZnO nanocomposite had the highest tensile strength and maximum bending resistance [10]. Topkaya et al., studied the effect of graphene nanoparticle reinforcement on the mechanical properties of glass fibre reinforced polymer composites, carbon fibre reinforced polymer composites and aramid fibre reinforced polymer composites. It was found that the tensile strength of the specimens increased with an increase in weight percentage of the graphene nanoparticle till a certain weight percentage after which the tensile strength started to decrease [11]. Su et al., fabricated carbon fibre reinforced polymers and basalt fibre reinforced polymers with various weight percentages of silica nanoparticles and found that the addition of silica nanoparticle improved the mechanical properties up to a certain weight percentage after which it started to decrease [12]. Mourad et al., fabricated Kevlar/Epoxy composites with various nanofillers such as alumina, silicon carbide, and multi-walled carbon nanotubes to study their effect on the various mechanical properties and observed that the addition of these nanofillers improved the mechanical properties of the Kevlar/Epoxy composite [13]. Suresh et al., investigated the effect of SiC nanoparticles on the mechanical properties of the Kevlar/Epoxy composites and concluded that the addition of the SiC nanoparticles enhanced the mechanical properties of the Kevlar/Epoxy composite until a certain weight percentage [14]. Mehmet Bulut et al., investigated the effect of SiC particles on the tensile and impact properties of Kevlar, carbon, and S-glass/epoxy composites and found that the addition of the SiC particles had a significant impact on the tensile and impact properties of the composite specimens [15]. Shubbar et al., investigated the influence of zinc oxide and titanium dioxide nanoparticles on Kevlar/Epoxy composites and found that the addition of the nanoparticles improved the mechanical properties of the Kevlar/Epoxy composite [16].

Researchers have established the benefits of using various nanofiller materials without combining two different nanofiller materials in assessing the mechanical properties of FRP composites. Not much published information is available to study the combined effect of two different nanofiller materials on the mechanical properties of fibre reinforced polymer composites. Literature review indicates that ZnO and SiC are used separately as nanofiller materials. It has been observed from the literature survey that there is no published information available containing ZnO and SiC nanofillers mixed together in the manufacturing of FRP composites. However, ZnO and SiC are independently used along with other fibre materials (Glass, Carbon, Kevlar) while fabricating FRP composites [9, 10, 13–16]. In this study, an attempt has been made to investigate the combined effect of silicon carbide and zinc oxide nanofillers on the mechanical properties of FRP composite comprising of Kevlar and epoxy.

Materials and Experimentation

The materials used for fabricating the specimens for this study are Kevlar Fibre Mat, epoxy resin with hardener, SiC nanoparticles, and ZnO nanoparticles. The Kevlar Fibre Mat was used as the reinforcement material, it was of 170 GSM supplied by Suntech Fibres Private Limited, Bangalore, Karnataka, India. The epoxy resin LY556 and the hardening agent HY951 were used as the matrix material and they were supplied by Zenith Industrial Supplies, Bangalore, Karnataka, India. The specifications of the resin and hardener are detailed in Tables 1 and 2, respectively. The resin mixture comprising of epoxy resin and hardener has been prepared in the ratio of 10:1. The ZnO nanoparticles were supplied by Adnano Technologies, Shimoga, Karnataka, India. The SiC nanopowder was supplied by Ultrananotech

Private Limited, Bangalore, Karnataka, India. Table 3 details the specifications of the nanoparticles.

Experimental work comprises three stages: (i) fabrication of the fibre reinforced polymer nanocomposite specimens, (ii) cutting of specimens as per ASTM standards (ASTM D3039, ASTM D790, and ASTM D2344) for testing using the abrasive waterjet cutting process, and (iii) testing of the specimens as per ASTM standards: tensile test (ASTM D3039), flexural test (ASTM D790), and Interlaminar Shear (ILS) strength test (ASTM D2344).

The specimens were fabricated by using hand layup method followed by vacuum bag molding. The Kevlar Fibre Mat was cut into 30 × 30 cm squares. The ZnO and SiC nanoparticles with different weight percentages (Table 4) were then stirred into the epoxy-hardener mixture for 10 min. The mixture was then spread on the worktable and a Kevlar Fibre Mat was placed on this layer. The epoxy-hardener-nanoparticle resin mixture was then poured onto the Kevlar Fibre Mat and evenly spread using a roller. This process was repeated until the required layers of mat were obtained. The layup layer orientation was kept at $0^0/90^0$ for all the layers. The specimen was then subject to vacuum bag molding (Fig. 1) to get a better finish, following which

Table 1 Specifications of the epoxy resin

Epoxy Resin—LY556	
Chemical name	Phenol 4,4-(-1-methyleneethylienedene) bipolymer with (chloromethyl) oxirane
State	Liquid
Color	Light yellow
Density	1.2 g/cm^2
Viscosity	1800–2200 MPa/s

Table 2 Specifications of the hardening agent

Hardening Agent—HY951	
Chemical name	Triethylenetetramine (TETA)
State	Liquid
Color	Clear pale yellow
Density	0.937 g/cm^2
Viscosity	10–20 MPa/s

Table 3 Specifications of the nanoparticles

Nanoparticle	Form	Purity	Color	Size	CAS No
Silicon carbide (SiC)	Powder	99.9%	Greenish grey	30–50 nm	409–21-2
Zinc oxide (ZnO)	Powder	99%	White	30–50 nm	1314–13-2

the specimen was cured in a hot air oven at 100^0C for an hour. One of the fabricated composite specimens is shown in Fig. 2.

The fabricated fibre reinforced polymer composites were then cut according to dimensions mentioned in ASTM standards for mechanical characterization. The specimens were cut by an abrasive waterjet machine (Maxiem 1515 from OMAX Corp., WA, USA) (Fig. 3) to reduce the risks of delamination and fibre pullout. The scope of the present work does not include characterization to check for delamination, voids, and other defects.

The drawings of the specimen prepared for conducting tensile strength, flexural strength, and interlaminar shear strength as per the respective ASTM standards are shown in Fig. 4.

The characterization of the fabricated specimens involved conducting of three tests, namely, tensile test (ASTM D3039), flexural test (ASTM D790), and interlaminar shear (ILS) strength test (ASTM D2344). The tests were performed using a Universal Testing Machine (Series 7200 from Dak Systems Inc., Maharashtra, India).

Table 4 Results of tensile strength, flexural strength, and interlaminar shear strength

Specimen no.	Wt. % of SiC	Wt% of ZnO	No. of layers	Tensile strength (MPa)	Flexural strength (MPa)	ILS strength (MPa)
1	0.5	0.75	8	289.247	207.501	79.539
2	0.5	1	9	311.25	229.069	83.563
3	**0.75**	**1**	**8**	**347.014**	**247.798**	**102.274**
4	0.75	0.75	9	296.559	208.683	74.971

Fig. 1 Vacuum bag molding process

Fig. 2 A fabricated composite specimen

Fig. 3 Abrasive waterjet cutting of fabricated specimens

Results and Discussion

The results of the tests conducted on the fibre reinforced polymer composite specimens using the Universal Testing Machine are given in Table 4.

It can be observed that specimen number 3, with 0.75% SiC, 1% ZnO, and 8 layers of Kevlar Fibre exhibited the best mechanical properties.

Table 5 compares the test results of the mechanical properties of the FRP specimen of present work with that of the properties of specimens taken from the literature

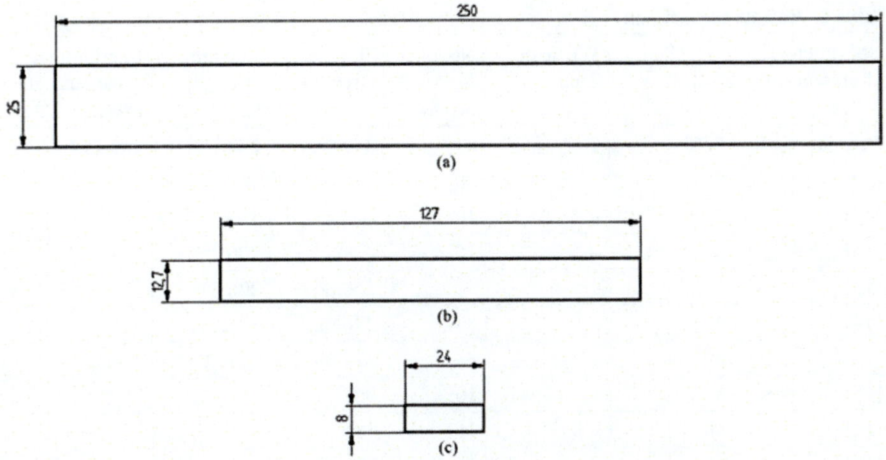

Fig. 4 Drawings showing the dimensions of specimens as per ASTM standards for: **a** Tensile test (ASTM D3039), **b** Flexural test (AST, D790), and **c** ILS test (ASTM D 2344)

survey. It can be observed that the fabricated specimen from the present work displays better properties than most of the specimens from literature survey. Only specimens from Reference [16] exhibited better flexural strength than the present work. This is due to the fact that the specimens in Reference [16] were structurally different when compared to the prepared specimen. It is also observed that the weight percent of the mixture of the two nanoparticles used in the present work is less than the weight percent of the nanoparticles of other specimens which are being compared with.

Conclusions and Future Scope

It can be concluded from the above studies that

- The composite specimen with the mixture of 0.75% and 1% weight percentages of SiC and ZnO nanoparticles, respectively, and made up of 8 layers of Kevlar Fibre Mat exhibited better mechanical properties.
- The weight percentage of SiC was the parameter which influenced the properties of the specimens the most.
- The addition of two nanofillers mixed together to the composite definitively enhances its mechanical properties when compared to adding just one nanofiller.
- As scope for future work, morphological and optimization studies for certain intended applications are planned.

Table 5 Results comparison

Specimen [References]	Wt. % of SiC	Wt. % of ZnO	No. of Kevlar layers	Tensile strength (MPa)	Flexural strength (MPa)	Interlaminar shear strength (MPa)
Present work	**0.75**	**1**	**8**	**347.014**	**247.798**	**102.274**
[13]	1	–	10	–	160	–
[13]	2	–	10	–	180	–
[13]	3	–	10	–	200	–
[14]	2	–	8	260.35	226.81	–
[14]	4	–	8	276.02	183.2	–
[14]	6	–	8	311.39	171.59	–
[15]	5	–	16	368	–	–
[15]	10	–	16	386	–	–
[15]	15	–	16	391	–	–
[15]	20	–	16	356	–	–
[16]	–	1	1	104.52	249.85	–
[16]	–	2	1	114.613	338.8	–
[16]	–	4	1	131.25	504	–

References

1. Koo JH (2006) Polymer nanocomposites: processing, characterization, and applications. McGraw Hill, New York
2. Paul DR, Robeson LM (2008) Polymer nanotechnology: nanocomposites. Polymer 49(15):3187–3204
3. Okpala CC (2013) Nanocomposites—an overview. IJERD 8(11):17–23
4. Sujesh G, Ganesan C (2014) Tensile behaviour of nano filled GFRP at different strain rates. IJMMME 2(2):93–96
5. Nayak RK, Mahato KK, Ray BC (2016) Water absorption behaviour, mechanical and thermal properties of nano TiO_2 enhanced glass Fibre reinforced polymer composites. Compos A Appl Sci Manuf 90:736–747
6. Raghavendra G, Ojha S, Acharya SK, Pal SK (2015) Influence of micro/nanofiller alumina on the mechanical behaviour of novel hybrid epoxy nanocomposites. High Perform Polym 27(3). https://doi.org/10.1177/0954008314550889
7. Panse P, Anand A, Murkute V, Ecka A, Harshe R (2014) Mechanical properties of hybrid structural composites reinforced with nanosilica. Polym Compos. https://doi.org/10.1002/pc.23286
8. Rajmohan T, Koundinya UK, Arun Premnath A, Harish G (2013) Evaluation of mechanical properties of nano filled glass Fibre reinforced composites. Paper presented at the international conference on advanced nanomaterials and emerging engineering technologies. Chennai, India
9. Naga Raju B, Ramji K, Prasad V (2015) Mechanical properties of glass fibre reinforced polyester ZnO nano composites. Mater Today: Proc 2(4–5):2817–2825. https://doi.org/10.1016/j.matpr.2015.07.294
10. Bose KJC, Thiagarajan A, Venkatesh DN, Velmurugan K (2019) Effects of ZnO nano reinforcements in the polymer matrix on the GFRP composites fabricated through VARTM. Mater Today: Proc 19(1). https://doi.org/10.1016/j.matpr.2019.08.081

11. Topkaya T, Çelik YH, Kilickap E (2020) Mechanical properties of fibre/graphene epoxy hybrid composites. J Mech Sci Technol 34:4589–4595. https://doi.org/10.1007/s12206-020-1016-4
12. Su C, Wang X, Ding L, Wu Z (2020) Enhancement of mechanical behavior of FRP composites modified by silica nanoparticles. Construct Build Mater 262. https://doi.org/10.1016/j.conbuildmat.2020.120769
13. Mourad AHI, Cherupurakal N, Hafeez F, Barsoum I, A Genena F, S Al Mansoori M, Al-Marzooqi LA (2020) Impact strengthening of laminated kevlar/epoxy composites by nanoparticle reinforcement. Polymers (Basel) 12(12):2814. https://doi.org/10.3390/polym12122814
14. Suresh A, Bhargavi P, Kiran Kumar M (2020) Simulation and mechanical characterization on kevlar epoxy reinforced composite with silicon carbide filler. Mater Today: Proc 38(5):2988–2995. https://doi.org/10.1016/j.matpr.2020.09.321
15. Bulut M, Alsaadi M, Erkliğ A (2018) A comparative study on the tensile and impact properties of Kevlar, carbon, and S-glass/epoxy composites reinforced with SiC particles. Mater Res Express 5(2). https://doi.org/10.1088/2053-1591/aaa991
16. Sawsan DA, Shubbar, Diwan MA, Kadhim AA, Diwan AA (2023) Influence of zinc oxide and titanium dioxide nanoparticles on kevlar/epoxy composites. J Compos Adv Mater 33. https://doi.org/10.18280/rcma.330304

Impact Evaluation of Corn Husk Reinforced Epoxy Composites

R. R. R. Junior, D. L. R. Oliveira, F. P. D. Lopes, N. T. Simonassi, C. M. F. Vieira, and S. N. Monteiro

Abstract Natural lignocellulosic fibers (NFLs) are considered a promising alternative to synthetic fibers in composite materials due to their economic, technical, environmental, and social advantages. In this study, corn stalk husk, an abundant and significant part of corn, was investigated as reinforcement in epoxy resin composites. The addition of 10, 20, and 30 vol% cornstalk husk was performed and the properties related to impact strength were evaluated in Izod and Charpy configurations, according to ASTM-D256 and ASTM-D6110 standards. Ten impact specimens were tested for each level of fiber reinforcement (vol%). The results revealed that the incorporation of 30 vol% cornstalk husk as reinforcement in the composites resulted in a significant increase in toughness, evidenced by Izod and Charpy impact test. All results were statistically analyzed by analysis of variance (ANOVA), with a 95% confidence level.

Keywords Natural fibers · Epoxy · Composites

Introduction

The technological issue and the need for new materials, something that has worried experts is the planetary environmental issue, which has been attacked by emissions of polluting gases and the incorrect disposal of non-degradable materials [1]. In contrast to the environmental crisis, we can also highlight the great demand for resources that has been increasing dramatically over time, according to Velasco, et al. [2] There is

R. R. R. Junior (✉) · D. L. R. Oliveira · F. P. D. Lopes · N. T. Simonassi · C. M. F. Vieira · S. N. Monteiro
LAMAV—Advanced Materials Laboratory, UENF—State University of the Rio de Janeiro, Av. Alberto Lamego, 2000, Campos Dos Goytacazes, Rio de Janeiro 28013-602, Brazil
e-mail: rogeriorabello30@gmail.com

F. P. D. Lopes
e-mail: perisse@uenf.br

C. M. F. Vieira
e-mail: vieira@uenf.br

B. Wisner et al. (eds.), *Composite Materials*, The Minerals, Metals & Materials Series,
https://doi.org/10.1007/978-3-031-50180-7_3

a need for a planet with approximately 75% more resources than already exist; in other words, human beings are theoretically using 1.75 planets. With this in mind, it is important to use sustainable materials.

A growing area of study and development of sustainable materials is focused on the manufacture of composite materials. This is due to its good characteristics that this class of materials has, some of which are: high specific strength, strong damping capacity, and high specific modulus [3]. A composite material is characterized by containing a matrix phase and a dispersed phase, also called reinforcement, seeking to combine the quality of the two applied phases, thus producing a material with flexible characteristics for various types of applications [4].

One of the most widely applied matrices in industry in various fields, such as automotive, military, aerospace, is the polymer matrix, this is due to the fact that the matrix has interesting characteristics for industrial applications, some of these characteristics are low cost in general, ductility, lightness, and good workability [5]. One of the matrix phases that is used extensively in industry is epoxy, a thermosetting polymer made up of interlocking chains that is unfortunately not biodegradable [6]. However, the reinforcements most commonly used in this matrix are synthetic materials, which further reinforces environmental problems because they generate a great deal of energy in their production and because they are abrasive to the machinery used in the production of their own, thus denoting the importance of using natural materials and waste in the manufacture of composite materials [7, 8].

One line of research being carried out at various institutions is the ballistic resistance of composites reinforced with natural materials [9], examples of work using natural materials to produce ballistic plates are the research by Filho et al. [10] which uses piassava fiber in the manufacture of ballistic plates, it is observed that it is possible to use it if there is an interfacial improvement between the fiber and the matrix and the work of Monteiro et al. [11] which uses curaua fiber and observes that the formulation with 30% curaua applied to an epoxy resin generates better resistance than the aramid fabric normally used in ballistic vests. This shows that synthetic fibers can be replaced by natural ones.

An example of a food that is produced worldwide on a large scale is corn; in the year 2021, there were about 8,858,138.91 tons of green corn produced in the world [12]. The country that stands out the most in this production is the United States. The chart below shows the 10 countries with the highest production between 1994 and 2021 (Fig. 1).

In addition to the corn produced, the stalk, the part that supports the plant, is often discarded or burned. Observing the large production of this grain, it can also be seen that there is a large production of this residue, thus denoting the need to use this by-product, which is produced on a large scale, in order to reduce energy, environmental, and climate impacts [13].

The aim of this work is to analyze the Charpy and Izod impact resistance of epoxy composites reinforced with corn husks in order to determine future tests for possible application in ballistic plates for personal defense.

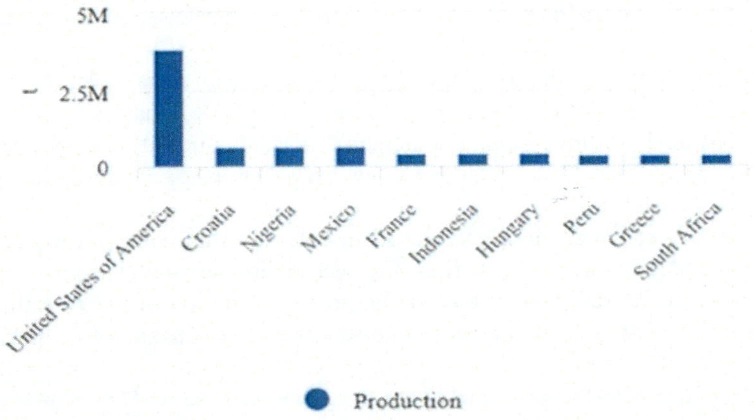

Fig. 1 Top ten productors of green corn maize. [12]

Materials and Methods

The corn husk fibers used in this work were donated by the Antônio Sarlo Agricultural School, located in Campos dos Goytacazes, Rio de Janeiro. After the fibers were extracted by hand (Fig. 2a), they were sanitized under running water in order to remove any impurities which could end up damaging the interface of the composite material. The average width defined for the corn husks was 4.50 ± 0.30 mm, Fig. 2b. The required fiber length is 150 mm and they are dried in an oven at 70 °C for 24 h or until their weight becomes constant, Fig. 2c.

The formulations evaluated in this work were developed using a diglycidyl ether bisphenol A (DGEBA)/triethylenetetramine (TETA) epoxy system, purchased from the AVIPOL company, located in São Paulo-SP, Fig. 2. A phr (parts of hardener per

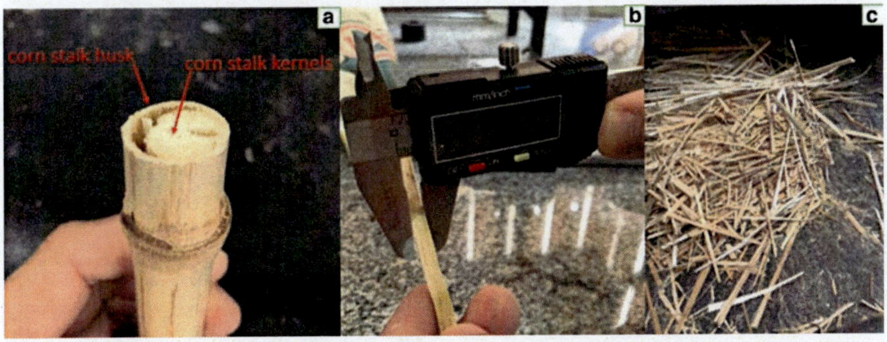

Fig. 2 **a** Corn husk and core, **b** width of dispersed phase, **c** material ready for the production of composites

resin) of 16% was used in the manufacture of the specimens used in this work, with a ratio of 16 parts of hardener to 100 parts of resin.

To manufacture the composite plates, a metal mold with rectangular dimensions of 150 × 120 × 12 mm^3 was used, using proportions of 10, 20, and 30% by volume of corn stalk husk. Information in the literature indicates a density of approximately 1.1 g/cm^3 for DGEBA [14]. Previous studies have established a density of 0.690 g/cm^3 for corn husk [15].

The fibers were meticulously arranged in layers in the mold, followed by the addition of a predetermined proportion of resin and hardener to fill the empty spaces. The resulting board was obtained through the pressing process in a hydraulic machine model Ribeiro RP-0002 under a constant load of 6 tons, maintained for a minimum of 24 h.

The Izod impact test was conducted in accordance with the ASTM-D256 standard [16] in order to measure the fracture energy of the composites in Joules per meter (J/m). Ten specimens were made with dimensions of 63.5 × 12.7 × 10 mm, as shown in Fig. 3b, with a notch at an angle of 45 ± 1° and a depth of 2.53 mm, as shown in Fig. 3a. The test was carried out using Pantec equipment (model XC.50) with an 11 J impact hammer, provided by the Universidade Estadual do Norte Fluminense (UENF).

Impact tests were carried out following the guidelines established by the ASTM-D6110 standard [17], using Charpy configurations. The tests were conducted at room temperature using a Pan 50 J impact testing machine developed by Panambra Técnica. Five samples of each material variant under investigation were examined. The results obtained were subjected to statistical analysis in order to assess the energy absorption

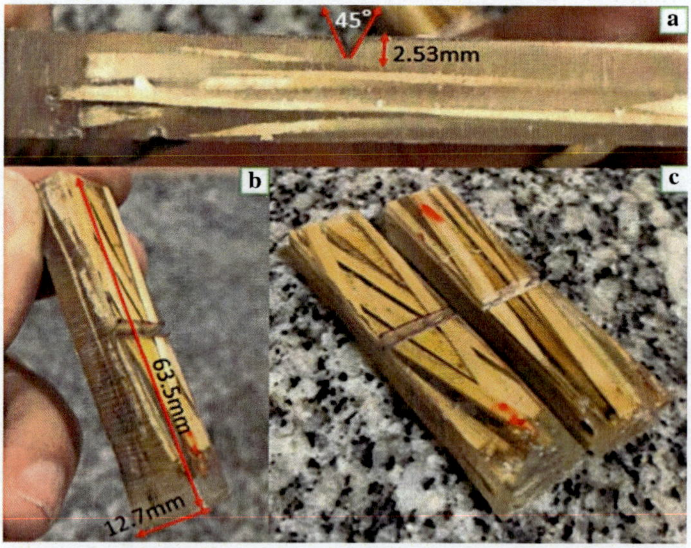

Fig. 3 **a** Notch size, **b** length and thickness, **c** specimens for Izod impact with 10% vol. samples

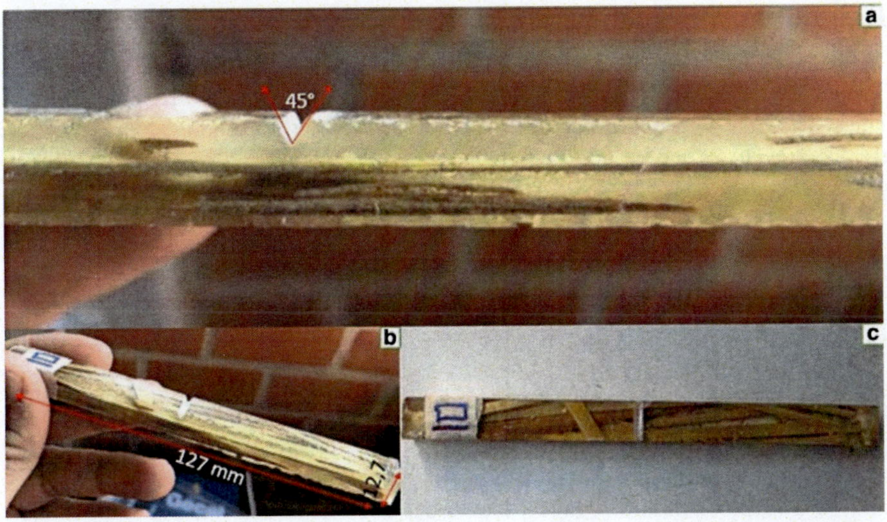

Fig. 4 **a** Notch measurement, **b** length and thickness, **c** Izod impact test specimens with 10% vol

capacity of the different materials. Six specimens were made with dimensions of 127 × 12.7 × 10 mm, as shown in Fig. 4b, with a notch at an angle of 45 ± 1° and a depth of 2.53 mm, with a distance between supports of 101.6 mm.

An analysis of variance (ANOVA) using the F-test was carried out in order to investigate possible significant differences between the means of the results obtained for the absorbed impact energy as the fiber volume fraction varied. All tests were conducted with a 95% confidence level.

The statistically significant differences between the averages of the results obtained for the various volume fractions of corn husk fibers were obtained and are presented in the results of this article. After obtaining this value, Tukey's test was applied at a 5% significance level. This test, widely recognized as a method for evaluating Least Significant Differences (LSD), aimed to quantitatively evaluate each of the fibre percentages used in the study. The Tukey test, an approach based on hypothesis testing, allows the hypothesis of equality to be rejected on the basis of the Honest Significant Difference (HSD), which is calculated as follows:

$$HSD = q \times \frac{\sqrt{(\text{MSE})}}{r}$$

In the equation presented, the variable denoting the total amplitude under investigation has a magnitude influenced by the degree of freedom (DF) of the residuals and the number of treatments used. The q tabulated is used to compare treatment averages. The mean square error, represented by MSE, plays an essential role in statistical formulation and analysis, while r refers to the number of repetitions carried out for each treatment [18].

Results and Discussion

The Charpy and Izod tests were carried out to see if there was a difference in resistance. The Charpy test, for example, is widely used in the automotive industry, while the Izod test is a good parameter for a possible future ballistic test.

Izod Impact Tests

The results of the Izod impact tests are shown in Table 1 and in Fig. 5. These analyses show a gradual increase in the range from 0 to 30% by volume of corn husk fibers incorporated into the epoxy matrix. However, there is a significantly more pronounced increase in the 20–30% range, characterized by a clear slope (energy absorbed per volume fraction of corn husk fibres). On the other hand, in the 0–10% and 10–20% fiber fraction ranges, the values are considerably lower. This notable variation in behavior can be attributed to a change in the fracture mode, which manifests itself in the 20–30% range by fiber volume. This phenomenon will be explored on the basis of macro-structural analysis subsequently.

Based on the results derived from the analysis of variance, as shown in Table 2, the hypothesis of equality of means was rejected at a significance level of 5%. This outcome was supported by the statistical test, which showed that the F value (calculated) exceeded the critical F value (tabulated). Consequently, it has been established that the variation in the volume fraction of corn stalk husk in epoxy matrix composites does indeed have a significant influence on the Izod impact energy.

Table 1 Results of the Izod impact test for epoxy matrix composites reinforced with corn stalk husks

Energy absorbed (J/m)—Izod test				
Samples	0%	10%	20%	30%
1	15.86	44.68	44.60	145.99
2	21.71	37.29	63.46	159.72
3	23.87	90.03	80.21	132.51
4	23.33	73.96	96.66	133.81
5	19.26	33.78	35.46	108.40
6	23.33	26.40	66.14	111.61
7	19.49	36.36	44.64	111.91
8	21.99	*	*	110.82
9	11.47	*	*	110.72
10	19.92	*	*	*
Average	20.02	48.93	61.60	125.05
Standard deviation	14.97	560.76	476.58	351.38

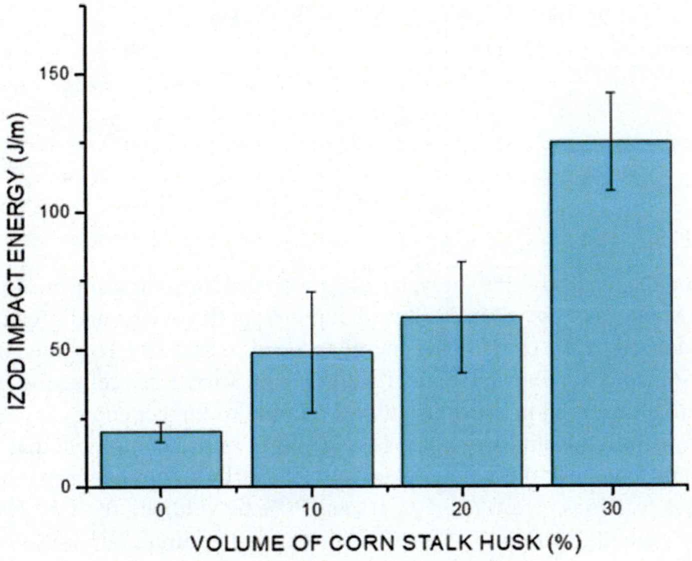

Fig. 5 Impact energy Izod X volumetric fractions of corn stalk husk

Table 2 Analysis of variance carried out on the means of the impact energies obtained from epoxy matrix composites reinforced with different percentages of corn stalk husk, ranging from 0 to 30% by volume

ANOVA

Variation causes	DF	Sum of squares	Average square residue	F (calculated)	P	F critical (tabulated)
Between groups	54,512,5025	3	18,170,83,415	57,46,586,659	2,56259E-12	2,93,402,989
Within groups	9169,86,416	29	316,2,022,124			
Total	63,682,3666	32				

In addition, Tukey's test was used to compare the means, using a 95% confidence level, in order to determine which of the corn stalk husk volume fractions results in superior performance in terms of Izod impact energy. The results of this analysis are shown in Tables 3 and 4

Table 3 Tukey test carried out to check the influence of the volume of thatch husk fibres on the Izod impact test

QMR	No. of treatments	Level of freedom	No. of repetitions	q-tabulated	DMS
316,2,022,124	4	29	8,25	3,72	23,030,238

Table 4 Results of the Tukey test carried out on the 0–30% samples

Volume fraction of corn stalk husk	0%	10%	20%	30%
0%	0	28,906,164	41,575,585	105,03,004
10%	28,906,164	0	12,669,421	76,123,881
20%	41,575,585	12,669,421	0	63,45,446
30%	105,03,004	76,123,881	63,45,446	0

Based on the results of this study, it can be seen that the composite reinforced with 30% corn stalk husk by volume showed superior performance and greater significance, evidenced by the Izod impact energy value of 125.05 J/m. However, the 10 and 20% fractions did not show statistically significant variations, reflecting behaviour similar to that observed in previous studies relating to impact energy in composites reinforced with lignocellulosic fibres [19, 20]. It is important to note that there is a significant difference in the average Izod impact energy values between the different mallow fibre fractions investigated (0, 10, and 30% by volume), with the 10 and 20% percentages showing similar values. This confirms the direct relationship between increasing fibre volume fraction and Izod impact energy performance.

Charpy Impact Test

The results of the Charpy impact tests are compiled in Table 5 and shown in Fig. 6. These analyses reveal a progressive increase in the range of 0–30% by volume of corn stalk husk fibres incorporated into the epoxy matrix. However, there was no substantial gain in terms of mechanical performance in the samples due to the presence of a high standard deviation. On the other hand, at the ends of the volume fraction range (0 and 30%), there were signs of significant changes in the fracture modulus.

Table 5 Charpy impact test results for epoxy matrix composites reinforced with corn stalk husks

Energy absorbed (J/m)—Charpy test					
Samples		0%	10%	20%	30%
1		22,39	72,07	107,14	102,22
2		32,23	37,28	53,33	87,57
3		57,51	35,71	65,62	103,05
4		27,65	62,56	44,17	81,41
5		36,42	62,61	44,29	114,44
6		22,44	52,96	*	128,21
Average		33,11	53,86	62,91	102,81
Standard deviation		173,18	217,78	688,40	294,21

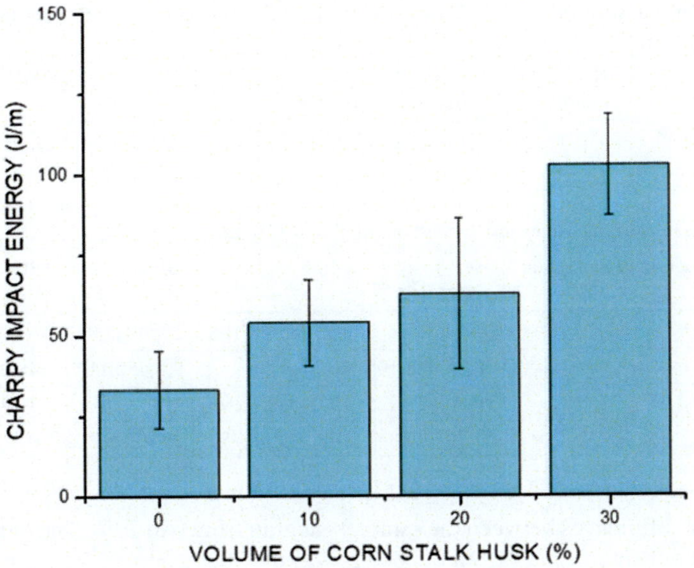

Fig. 6 Charpy impact energy X volume fractions of corn stalk husk

Following the results obtained from the analysis of variance, as shown in Table 6, the hypothesis of equality of means was rejected at a significant level of 5%. This result was confirmed by the statistical test, in which the calculated F value exceeded the established critical F value. It can therefore be said that the variation in the volume fraction of corn stalk husk in epoxy matrix composites has a significant influence on the Charpy impact energy. Therefore, the Tukey test will indicate which group suffers statistically significant variations, with a significant level of 95%, as shown in Tables 7 and 8.

Analysis of the results shows that, in most of the comparisons made, the value of the Minimum Significant Difference (MSD) predominated, indicating the absence of

Table 6 Analysis of variance performed on the mean Charpy impact energies obtained from epoxy matrix composites reinforced with different percentages of corn stalk husk, ranging from 0 to 30% by volume

ANOVA						
Variation causes	DF	Sum of squares	Average square residue	F (calculated)	P	F critical (tabulated)
Between groups	15,372,7286	3	5124,242,852	15,7,555,447	2,18005E-05	3,127,350,005
Within groups	6179,45,086	19	325,2,342,556			
Total	21,552,1794	22				

Table 7 Tukey test carried out to check the influence of the volume of thatch husk fibers on the Charpy impact test

QMR	No. of treatments	Level of freedom	No. of repetitions	q-tabulated	DMS
325,2,342,556	4	19	5,75	3,98	29,9,327,963

Table 8 Results of the Tukey test carried out on the 0–30% samples

Volume fraction of corn stalk husk	0%	10%	20%	30%
0%	0	20,75,928,105	29,8,042,922	69,7,079,684
10%	20,75,928,105	0	9,0,450,112	48,9,486,874
20%	29,80,429,225	9,045,011,199	0	39,9,036,762
30%	69,70,796,841	48,94,868,736	39,9,036,762	0

significant differences between the samples ranging from 0 to 20%. The only sample that stood out significantly from the others was the 30% sample.

Conclusions

The results obtained in this study show that

- The 30% sample shows better energy absorption capacity in the Izod and Charpy impact tests.
- Brittle fracture was noted.
- The results expected in this work coincided with studies found in the literature.

Acknowledgements The authors would like to thank the funding agencies CAPES, CNPQ (process no. 302976/2022-1), and FAPERJ (process nos. E-26/205.832/2022, E-26/203.874/2022, E-26/200.847/2021, and E-26/201.628/2021).

References

1. Madueke CI, Mbah OM, Umanukwe R (2023) A review on the limitations of natural fibres and natural fibre composites with emphasis on tensile strength using coir as a case study. Polym Bull 80:3489–3506. https://doi.org/10.1007/s00289-022-04241-y
2. Velasco DCR, Lopes FPD, Souza D, Lopera HAC, Monteiro SN, Vieira CMF (2023) Evaluation of Composites reinforced by processed and unprocessed coconut husk powder. Polymers 15(5):1195. https://doi.org/10.3390/polym15051195
3. Kerni L, Singh S, Patnaik A, Kumar N (2020) A review on natural fiber reinforced composites. Mater Today: Proc 28:1616–1621. https://doi.org/10.1016/j.matpr.2020.04.851

4. Girijappa YGT, Rangappa SM, Parameswaranpillai J, Siengchin S (2019). Natural fibers as sustainable and renewable resource for development of eco-friendly composites: a comprehensive review
5. Yao J, Zhou Z, Zhou H (2019) Highway engineering composite material and its application. Springer, Berlin/Heidelberg, Germany
6. Oliveira DLR, Freitas LFF, Velasco DCR, Vieira CMF, Monteiro SN, et al (2023) Avaliação em compressão de compósitos epoxídicos reforçados com particulados finos de EUCALIPTO. Congresso Anual da ABM - Internacional, São Paulo. 76(76):2094–2107. https://doi.org/10.5151/2594-5327-40228
7. Colò F, Bella F, Nair JR, Destro M, Gerbaldi C (2015) Cellulosc-based novel hybrid polymer electrolytes for green and efcient Na-ion batteries. Electrochim Acta 174:185–190
8. Aziz SB, Abidin ZHZ (2013) Electrical conduction mechanism in solid polymer electrolytes: new concepts to Arrhenius equation. J Soft Matter. 1–8. https://doi.org/10.1155/2013/323868
9. Nurazzi NM, Asyraf MRM, Khalina A, Abdullah N, Aisyah HA, Rafiqah SA et al (2021) A review on natural fiber reinforced polymer composite for bullet proof and ballistic applications. Polymers 13(4):646. https://doi.org/10.3390/polym13040646
10. Garcia Filho FC, Luz FS, Oliveira MS, Bezerra WBA, Barbosa JDV, Monteiro SN (2021) Influence of rigid Brazilian natural fiber arrangements in polymer composites: energy absorption and ballistic efficiency. J Compos Sci 5:201. https://doi.org/10.3390/jcs5080201
11. Monteiro SN, Louro LHL, Trindade W, Elias CN, Ferreira CL et al (2015) Natural curaua fiber-reinforced composites in multilayered ballistic armor. Metall Mater Trans A 46:4567–4577. https://doi.org/10.1007/s11661-015-3032-z
12. Food and Agriculture Organizatio of the United Nations (2023) Crops and livestock products. https://www.fao.org/faostat/en/#data/QCL/visualize Acessed 14 Aug 2023
13. Al-Hashimi A, Evans GJ, Cox B (1994) Contaminants mobility in acid-generating wastes/subsoil systems. J Environ Sci Health Part A 29(4):745–752. https://doi.org/10.1080/10934529409376069
14. Silva TT, Silveira PHPM, Figueiredo ABHS, Monteiro SN, Ribeiro MP et al (2022) Dynamic mechanical analysis and ballistic performance of kenaf fiber-reinforced epoxy composites. Polymers 14(17):3629. https://doi.org/10.3390/polym14173629
15. Junior RRR, Ribeiro AM, Guimarães RS, Simonassi NT, Lopes FPD, et al (2023) Análise metodológica de confecção de compósitos de poliuretano reforçado por colmo DE MILHO. Congresso Anual da ABM—Internacional, São Paulo 76(76):2146–2158. https://doi.org/10.5151/2594-5327-40255
16. American Society of Testing and Materials, D256-10 (2018) Standard test methods for determining the izod pendulum impact resistance of plastics. West Conshohocken, PA, United States, ASTM international. https://doi.org/10.1520/D0256-23
17. American Society of Testing and Materials, D6110 (2018) Standard test method for determining the charpy impact resistance of notched specimens of plastics. West Conshohocken, PA, United States, ASTM international. https://doi.org/10.1520/D6110-18
18. Larson MG (2008) Analysis of variance. Circulation 117(1):115–121
19. Neuba LM, Junio RFP, Ribeiro MP, Souza AT, Lima ES, et al (2020) Promising mechanical, thermal, and ballistic properties of novel epoxy composites reinforced with cyperus malaccensis sedge fiber. Polymers 12(8):1776. https://doi.org/10.3390/polym12081776
20. Monteiro SN, Lopes FPD, Nascimento DCO, Ferreira AS, Satyanarayana KG (2013) Processing and properties of continuous and aligned curaua fibers incorporated polyester composites. J Market Res 2(1):2–9. https://doi.org/10.1016/j.jmrt.2013.03.006

Study of the Impact Behavior of Epoxy Matrix Composites with Granite Waste

J. A. T. Linhares Júnior, D. C. R. Velasco, F. P. D. Lopes, C. M. Vieira, A. R. G. Azevedo, and M. T. Marvila

Abstract The rock production process generates a significant amount of waste that can be utilized for various applications, such as the production of new composites. This allows for the acquisition of new materials while minimizing waste disposal and resource consumption. In this regard, the objective of this study is to evaluate how the incorporation of granite waste into epoxy matrix particulate composites affects impact resistance. The waste was characterized in terms of size as well as used to fabricate test specimens with 0, 12.5, 25, 37.5, and 50% waste (% vol.). These specimens were subjected to impact testing using the Izod configuration, following ASTM D256 standards. Finally, the results were analyzed using Analysis of Variance (ANOVA) and Tukey's test. These results allowed for the characterization of the waste and indicated that its inclusion significantly enhances impact resistance.

Keywords Composite · Waste · Sustainability

Introduction

Brazil assumes a crucial position as a producer and exporter of ornamental stones, effectively meeting the needs of several countries, as illustrated in Fig. 1. In this productive context, a lot of solid waste is generated at all stages of the production

J. A. T. L. Júnior · D. C. R. Velasco (✉) · F. P. D. Lopes · C. M. Vieira · A. R. G. Azevedo (✉)
LAMAV—Advanced Materials Laboratory, UENF—State University of the Northern Rio de Janeiro, Av. Alberto Lamego, 2000, Campos Dos Goytacazes, Rio De Janeiro 28013-602, Brazil
e-mail: davidc.r.v2014@gmail.com

A. R. G. Azevedo
e-mail: afonso@uenf.br

C. M. Vieira
e-mail: vieira@uenf.br

M. T. Marvila
UFV—Federal University of Viçosa, CRP—Rio Paranaíba campus; Rodovia BR 230 KM 7, Rio Paranaíba—MG 38810-000, Brazil
e-mail: markssuel.marvila@ufv.br

B. Wisner et al. (eds.), *Composite Materials*, The Minerals, Metals & Materials Series,
https://doi.org/10.1007/978-3-031-50180-7_4

line, which can become an environmental problem in cases of unnecessary disposal [1]. In general, ornamental stones such as granite go through several stages such as mining, cutting, calibration, grinding, and polishing. Waste in the form of solid particles generated in these steps can pollute the air, contaminate groundwater, and reduce soil quality after long periods, due to leaching [2, 3]. There is a need for studies aimed at the technological application of this waste.

One of the alternatives used to mitigate the problems generated by the large amount of waste resulting from the productive sector is its use in the preparation of other materials, such as epoxy matrix compounds. This resin has been standing out and gaining space in the academic environment because it is a thermosetting material with good mechanical, thermal, and chemical properties, in addition to being resistant to corrosive environments [5, 6].

Several studies have been carried out regarding the use of granite waste as particulate matter in epoxy resin systems. In general, the mechanical properties of composites are evaluated, but there are other investigations into technological properties and specific applications [7–9].

Arumugam et al. [10] used granite residue in epoxy matrices to evaluate its application as a high-voltage electrical insulator and the study concluded that the materials obtained satisfactory performance and could be used in the manufacture of various electrical insulation products. Rout e Satapathy [11] applied the granite residue next to the glass residue, as filler in an epoxy matrix in order to increase the mechanical properties, chemical resistance, and abrasion of the composites. The research found an increase in hardness, impact energy and resistance to erosion; however, there was

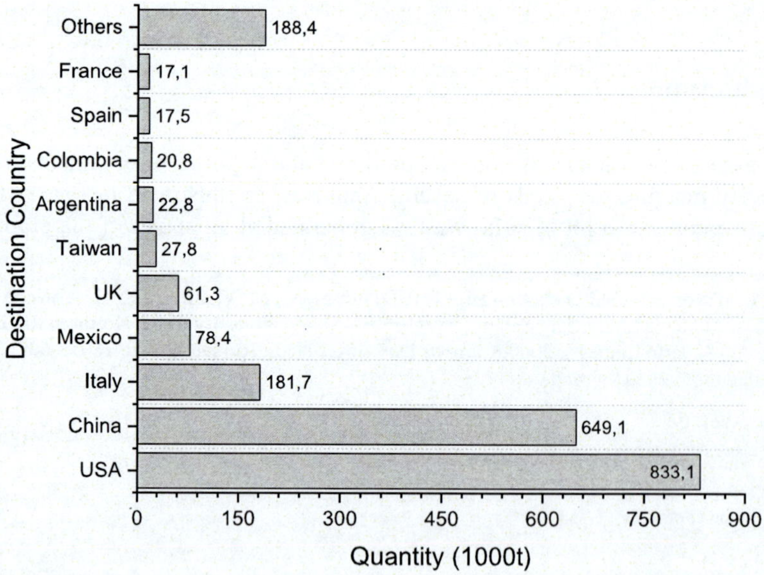

Fig. 1 Brazilian export of natural stones [4]

a decline in flexural strength, requiring a careful analysis before applying certain levels of granite waste.

In the context of the application of granite waste in epoxy matrix composites, this study aims to investigate the impact behavior of composites with the incorporation of granite particulate in different volumes. The particulate was characterized in relation to its particle size. Altogether five formulations were tested in impact tests (Izod), with 0, 12.5, 25, 37.5, and 50% particulate matter. Analysis of variance (ANOVA) and the Tukey test were also performed.

Materials and Methods

Initially, the granite particulate was sieved at 100 mesh and then characterized in terms of granulometry, using a laser granulometer, Microtrac S2500. Subsequently, the formulations with 0, 12.5, 25, 37.5, and 50% of particulates were prepared, with a stoichiometric proportion of 20 phr, using the system epoxy resin, SQ 2005/SQ 3131, composed of bisphenol A diglycidyl ether (DGEBA) with diethylene triamine hardener (DETA).

The specimens were made in silicone molds, measuring 10 × 12.5 × 60 mm. The cure used was 5 days at room temperature. After the curing process, the specimens were subjected to impact resistance tests, Izod. The ASTM D256-23 standard [12] provided the guidelines for the tests, which were performed on a PANTEC Pendulum impact testing machine, model XC-50, as shown in Fig. 2.

To evaluate the statistical difference between the groups of samples, analysis of variance (ANOVA) and the Tukey test were performed. A 5% significance level was adopted. The software used was PAST, which follows the Copenhagen and Holland algorithm [13].

Results and Discussion

The granite wastes were submitted to a laser granulometry analysis, resulting in the graphical representation of the cumulative particle size distribution, shown in Fig. 3. From this analysis, it was possible to determine an average size of 29.15 microns. In addition, an excellent fit of the data to a three-parameter Weibull distribution was observed, with a coefficient of determination (R^2) of 0.99, indicating a highly significant correspondence between this and the results of laser granulometry.

About the impact resistance, Fig. 4a can be generated, which presents the impact resistance, and Fig. 4b, which shows the results of the notch resistance. Through these, the incorporation of granite residues resulted in significant strength gains in all formulations, according to ANOVA. Regarding the formulation with the highest average resistance (with 37.5% of particulates), no significant resistance variation was observed in comparison with the formulations with 25 and 50%. This indicates

Fig. 2 PANTEC pendulum impact testing machine

the existence of a range of applicability for waste with a relatively constant energy absorption, without significant variations.

This behavior can be explained by the incorporation of particles, which facilitate the introduction of bubbles in the composite, which, in turn, tend to deflect the cracks. In addition, the rigid nature of granite can also contribute to crack deflection. [14, 15]. This pattern of behavior, with gains in resistance to impact with the incorporation of granite residues, has already been identified by Subhash et al. [16] in your work.

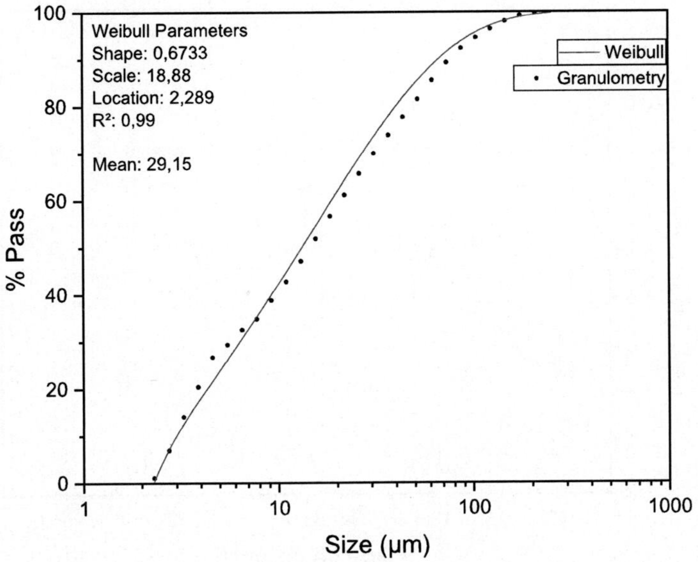

Fig. 3 Gypsum particulate cumulative distribution curve

Conclusions

Through this work, the residue can be characterized, observing an average particle size of 29.15 microns, as well as a remarkable fit to the parameters of the Weibull distribution. In addition, the incorporation of these residues in the composites resulted in significant improvements in impact resistance, with the incorporation of particulates in the range of 25–50% being around 30% more resistant than the matrix.

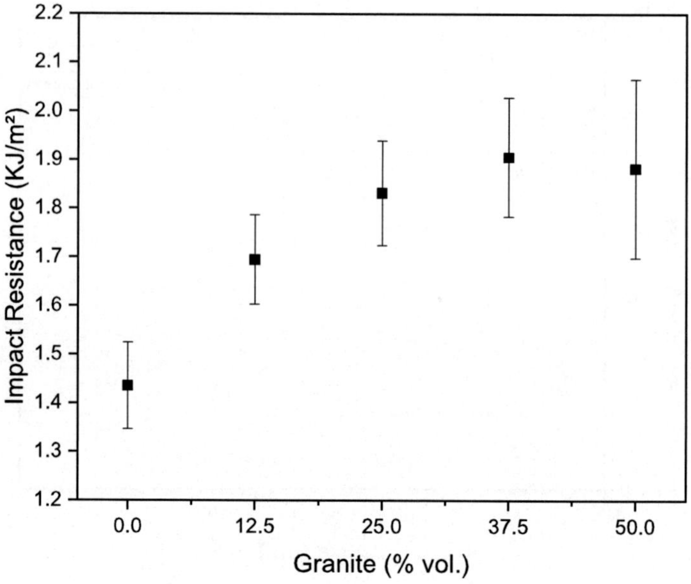

(a)

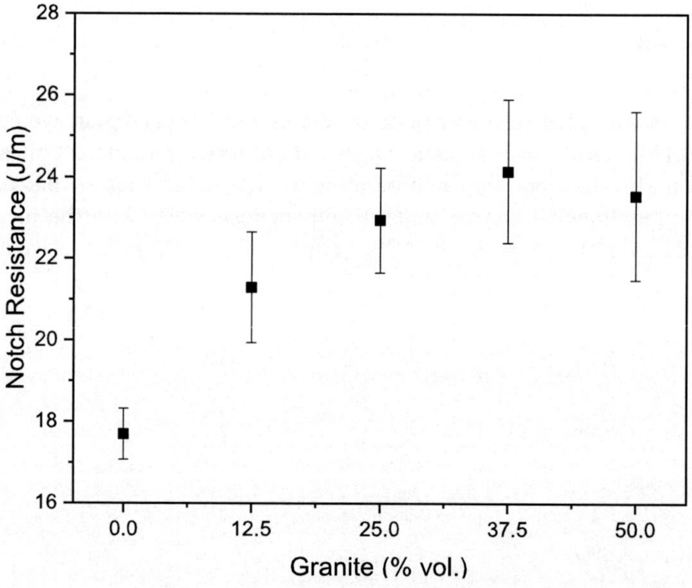

(b)

Fig. 4 Impact test results: **a** impact strength; **b** notch resistance

References

1. Silva TLC, Carvalho EAS, Barreto GNS, Silva TBP, Demartini TJC, Vieira CMF (2023) Characterization of artificial stone developed with granite waste and glass waste in epoxy matrix. J Market Res 26:2528–2538. https://doi.org/10.1016/j.jmrt.2023.08.045
2. Chaturvedi R, Asokan P, Prashant T, Ravi P, Anam K, Alka M, Manoj Kumar G, Kumar TV (2022) Next-generation high-performance sustainable hybrid composite materials from silica-rich granite waste particulates and jute textile fibres in epoxy resin. Ind Crops Prod 177. https://doi.org/10.1016/j.indcrop.2022.114527
3. Nayak SK, Satapathy A, Mantry S (2022) Use of waste marble and granite dust in structural applications: a review. J Build Eng 46:103742. https://doi.org/10.1016/j.jobe.2021.103742
4. Associação Brasileira da Indústria de Rochas Ornamentais (2022) Balanço das exportações e importações de rochas ornamentais em 2022. http://comexstat.mdic.gov.br
5. Gomes MLP, Carvalho EAS, Demartini TJC, de Carvalho EA, Colorado HA, Vieira CMF (2021) Mechanical and physical investigation of an artificial stone produced with granite residue and epoxy resin. J Compos Mater 55:1247–1254. https://doi.org/10.1177/0021998320968137
6. Maluga R, Kumar MS, Pati R, Kumar SS (2023) Physical, mechanical and wear characterization of epoxy composites reinforced with granite/marble powder. Mater Today Proc. https://doi.org/10.1016/j.matpr.2023.02.045.
7. Kampa Ł, Chowaniec A, Królicka A, Sadowski L (2022) Adhesive properties of an epoxy resin bonding agent modified with waste granite powder. J Coat Technol Res 19:1303–1316. https://doi.org/10.1007/s11998-022-00620-2
8. Chowaniec-Michalak A, Czarnecki S, Sadowski Ł (2022) Waste granite powders as fillers in epoxy coatings: a case study in a car repair workshop. Proc Inst Mech Eng Part C: J Mech Eng Sci. https://doi.org/10.1177/09544062221126639
9. Krzywiński K, Sadowski Ł, Fedoruk K, Sieradzki A (2023) Thermal properties of polymer floor coating with alternative granite powder filler. In: Chastre C et al (eds) Testing and experimentation in civil engineering. TEST&E 2022. RILEM Bookseries, vol 41. Springer, Cham. https://doi.org/10.1007/978-3-031-29191-3_19
10. Arumugam H, Iqbal M, Ahn CH, Rimdusit S, Muthukaruppan A (2023) Development of high performance granite fine fly dust particle reinforced epoxy composites: structure, thermal, mechanical, surface and high voltage breakdown strength properties. J Market Res 24:2795–2811. https://doi.org/10.1016/j.jmrt.2023.03.199
11. Rout AK, Satapathy A (2015) Study on mechanical and erosion wear performance of granite filled glass-epoxy hybrid composites. In: Proceedings of the institution of mechanical engineers, part l: journal of materials: design and applications. SAGE Publications Ltd, pp 38–50. https://doi.org/10.1177/1464420713499483
12. American Society for Testing and Materials (2023) D256: Standard Test Methods for Determining the Izod Pendulum Impact Resistance of Plastics, West Conshohocken. https://doi.org/10.1520/D0256-23E01
13. Holland B (1988) Computation of the distribution of the maximum studentized range statistic with application to multiple significance testing of simple effects. J Stat Comput Simul 30:1–15. https://doi.org/10.1080/00949658808811082
14. Velasco DCR, Lopes FPD, Souza D, Lopera HAC, Monteiro SN, Vieira CMF (2023) Evaluation of composites reinforced by processed and unprocessed coconut husk powder. Polymers (Basel) 15:1195. https://doi.org/10.3390/polym15051195
15. Mousavi SR, Estaji S, Paydayesh A, Arjmand M, Jafari SH, Nouranian S, Khonakdar HA (2022) A review of recent progress in improving the fracture toughness of epoxy-based composites using carbonaceous nanofillers. Polym Compos 43:1871–1886. https://doi.org/10.1002/pc.26518
16. Subhash C, Krishna MR, Raj MS, Sai BH, Rao SR (2018) Development of granite powder reinforced epoxy composites. Mater Today Proc 5:13010–13014. https://doi.org/10.1016/j.matpr.2018.02.286

Part II
Eco-Friendly and Sustainable Composite Materials: Waste Stream Benefits

Nanocomposite Materials for Radionuclide Sequestration from Groundwater Environments

Simona E. Hunyadi Murph

Abstract The half-lives of radionuclides range from fractions of a second to billions of years. Since no practical method of altering radioactive decay exists, and since exposure to either the energy emitted from radioactive decay or chemical properties of radionuclides poses dire health risks, radioactive materials must be segregated and controlled. The capture, treatment, and disposition of radioactive materials remain an extraordinary challenge. In here, we focus our attention on the synthesis and characterization of a unique class of nanocomposite materials that have potential for removal of radionuclide contamination. Specifically, we report a simple approach to decorate the surface of iron-based (Fe/Fe_xO_y) material with various nano-catalysts. Specifically, copper (Cu), tin (Sn), and silver (Ag) nanoparticles were prepared through two different reduction approaches, namely, citrate and cetyltrimethylammonium bromide (CTAB) methods, on the iron-based material surface. All samples were characterized by a variety of analytical tools, which included scanning electron microscopy (SEM), electron-dispersive X-ray microanalysis (EDS), and EDS mapping to elucidate materials' morphology as well as nano-catalysts' loading and location on the iron-based structures.

Keywords Nanocomposite · Radionuclides · Bimetallic compounds · Sequestration

Introduction

The radioactive nuclides have half-lives ranging from fractions of a second to minutes, hours or days, through to billions of years [1]. Since no practical method of altering radioactive decay exists, and since exposure to either the energy emitted from radioactive decay or chemical properties of radionuclides poses dire health

S. E. Hunyadi Murph (✉)
Savannah River National Laboratory, Aiken, SC 29808, USA
e-mail: Simona.Murph@srnl.doe.gov

University of Georgia, Athens, GA 30602, USA

B. Wisner et al. (eds.), *Composite Materials*, The Minerals, Metals & Materials Series,
https://doi.org/10.1007/978-3-031-50180-7_5

risks, radioactive materials must be segregated and controlled. The nature, volume, and magnitude of radioactive materials are vast and diverse. For example, at the Savannah River Site (SRS), during nuclear material production operations, over 37 million gallons of high-level nuclear waste (HLW), including mixtures of radioactive technetium, cesium, iodine, uranium, plutonium, strontium, etc., were generated as a byproduct of nuclear weapons production [2]. As expected, the stewardship, disposition, and environmental cleanup of nuclear materials are not trivial. It requires sophisticated and operationally complex materials and technologies to effectively isolate them from the environment. Improper treatment and handling would result in significant consequences to the environment [3].

Understanding the intricacies of the radionuclides fate, transport, and performance in the ecosystem is critical for the development of efficient strategies to protect the environment and human health. Therefore, the greatest problems associated with nuclear waste are related to the efficient capture, long-term storage, and disposal of the waste in a non-toxic form [4].

Almost all nuclear operations use sorption technologies to reduce waste volume or to recover key valuable elements or isotopes [5, 6]. Physical or chemical approaches are largely employed for removal of the radioactive elements from waste solutions through the use of adsorbents, absorbents, and ion exchange materials [7]. Technetium-99, a fission product of uranium-238 and a beta emitter, is a major contaminant at nuclear power plants that has been unintentionally released in the environment [8]. Zero valent iron was efficiently used as an effective remediation agent for radioactive Tc decontamination [6, 9]. A micelle directing surfactant, such as cetyltrimethylammonium bromide, was also used for capturing Tc moieties (TcO_4^-) [10]. Moreover, monosodium titanate material is currently used as a baseline sorbent for the removal of other radioactive contaminants, such as Sr-90 and alpha-emitting radionuclides at SRS [7].

Significant challenges remain, however, in the development of deployable techniques with high sensitivity and selectivity for the efficient treatment of radionuclides. This is due to the complex and challenging environments in which these technologies must operate, i.e. high/low pH, temperature/pressure fluctuations, radionuclide mixtures, solvents, neutralizing media, etc. Therefore, in order to accelerate the cleanup efforts, novel materials and technologies are still needed. Among the many classes of candidate materials, nanocomposite materials possess unique properties that could address several limitations and lifecycle schedules related to the nuclear waste materials processing [11, 12]. Nanomaterials' distinctive properties arise from their improved physical and chemical properties imparted over their single-component counterparts [13, 14]. Coupling materials with disparate functionalities and distinct properties into a single, hybrid/multifunctional operating material open the door to a myriad of possibilities and applications for enhanced detection, imaging, manipulation, encapsulation, etc. [15]. Additionally, the high surface area, ease of surface engineering, and diverse likelihoods for interfacial modifications make them highly sought for cleanup purposes [16]. As an example, we developed novel nanocomposite materials in the form of gold-monosodium titanate that not only capture the radioactive nuclides, but also determine the concentration of the radioactive sorbate in both solution and while captured onto the monosodium titanate solids. This innovative nanotechnology is deployable, cost-effective, and doesn't generate additional waste. It diminishes potential radiation exposure and reduces processing time for the sample preparation and analysis [7, 17].

In here, we focus our attention on the synthesis and characterization of a unique class of nanocomposite materials that could be used for the removal of radionuclide contaminants. Specifically, copper (Cu), tin (Sn), and silver (Ag) nanoparticles were prepared through two different reduction approaches, namely, citrate and cetyltrimethylammonium bromide (CTAB) methods, on the iron-based (Fe/FexOy) material. All samples were characterized by a variety of analytical tools, which included scanning electron microscopy (SEM), electron-dispersive X-ray microanalysis (EDS), and EDS mapping to elucidate materials' morphology as well as nano-catalysts' loading and location on the iron-based structures.

Experimental Details

Materials: All chemicals were purchased from Sigma-Aldrich. Porous iron material was provided by Höganäs Environmental Solutions (Cary, NC, USA). All glasswares used in the following protocols were cleaned with aqua regia and then rinsed with deionized water. A neodymium magnet was used to separate magnetic materials from aqueous solutions.

Fabrication and Characterization: The iron-based material was used as a template for creation of composite material by a straightforward wet chemical method that is amenable for scaling up. All nanoparticles were produced by reduction approaches as previously reported by us while using iron-based material as supports. Basically, metal ions were reduced by either sodium citrate or sodium borohydride in the presence of surfactants or reducing/capping reagents, such as cetyltrimethylammonium bromide (CTAB) or sodium citrate. In the case of the citrate reduction synthesis approach, 1.25×10^{-4} M metal ions was heated to boiling and 1 wt % reducing agent (citrate) solution was added in the presence of 0.1 g of iron-based material. In the case of the CTAB approach, 0.1 g of iron-based material was incubated in an aqueous solution of CTA with a 0.1 M concentration. Next, metal salt 250ul of 0.01 M were introduced into the flask while stirring at room temperature for 15 min. Under vigorous stirring, 600ul of sodium borohydride 0.01 M was added to this solution. Stirring continued for an additional 15 min. Samples were washed after 24 h and dried. Dried samples were placed on copper double tape and imaged with a Hitachi SU8200 Scanning Electron Microscope (SEM) coupled with Energy-Dispersive X-Ray Spectroscopy (EDS) to evaluate nanomaterial morphologies and compositions.

Results and Discussion

Characterization of Iron-Based Materials

Iron-based materials were used here as a template for creation of bimetallic catalytic materials with tailored and tunable structural, optical, and surface properties. Typically, zero valent iron material oxidizes upon oxygen and/or water exposure. Therefore, while not investigated in this study, it is postulated that the template material is a mixture of zero valent iron and iron oxide (Fe/Fe_xO_y) material. It was reported that the presence of hydroxyl radical, superoxide radical, and ferryl ion species enhances material's reactivity [18]. Consequently, this is beneficial for water treatment.

The template, i.e. Fe/FexOy material, was sonicated in either water or sodium citrate before decoration with metallic nanoparticles. Upon 10–15-min sonication, no significant changes were recorded to the material's pore structures and morphology. The material displays unique morphologies with irregular tubular structures and various pore-like and "hollow" environments. The tubular structures have variable configurations and geometries with the diameter ranging from around 1–4 um in size. The pore structure is of particular interest as it is beneficial during treatment of liquid nuclear waste (Fig. 1). Nitrogen sorption isotherm indicated that Fe/FexOy material has a BET surface area of 0.95 m^2/g, pore volume of 0.0068 mL/g, and an average pore diameter of 364 Å [4]. Energy-dispersive X-ray analysis (EDS) shows that over 90% of the material contains elemental Fe with other small percentage of impurities or support elements, i.e. Al, C.

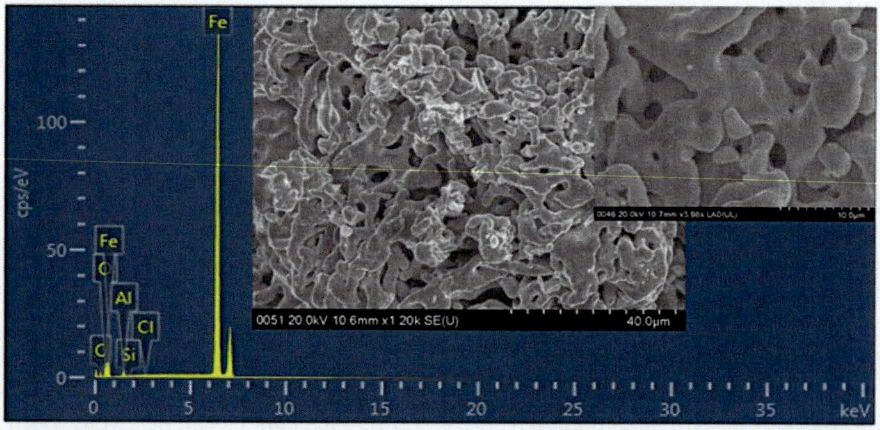

Fig. 1 SEM and EDS data collected on porous iron support materials

Production and Characterization of Nanoscale Decorated Fe/ FexOy Materials

Addition of a second metal onto the Fe/FexOy material, especially at the nanoscale, offers significant benefits for water remediation developments. For example, the presence of a nanoscale metal boosts material's reactivity though increased reductive properties and/or catalytic capabilities [19, 20]. These enhanced characteristics were successfully exploited for cleanup efforts, namely, sequestration and decomposition of organic and heavy metal contaminates [21–23]. The presence of a second metal, especially at the nanoscale, facilitates the efficient flow of electron transfer between iron and contaminants. Bimetallic materials also open the door to selectivity through tailored surface functionalization. For example, one could promote targeted sequestration pathways by using various bifunctional linkers for specific interactions, i.e. positive/negative surface charge, hydrophilic/hydrophobic surfaces, etc. Nanomaterials are of particular interests as sequestering agents due to their favorable properties comparing with the bulk materials [24, 25]. Nanoscale materials have an increased surface area when compared to counter bulk materials [26]. A significant portion of atoms are exposed to the environment and available for reactions. Therefore, enhanced reaction kinetics can be obtained by simply using nanoscale materials. Moreover, the surface atoms in nanomaterials have higher energy (due to under-coordination) than the bulk atoms making them highly reactive. Therefore, nanomaterials have higher remediation capacities than the bulk nanomaterials. Nanomaterials also serve as efficient catalysts/photocatalysts or storage media to boost chemical reactions [27]. Either through manipulation of size, shape, and/or composition, nanoscale material's properties can be tailored to drive specific and/or enhanced photo-catalytic or thermo-catalytic reactions [28, 29]. For example, titania nanoparticles are active only in the UV region of the spectrum [30]. By coupling titania to a second component, gold nanoparticles, the photo-catalytic response is enhanced from the UV to the Vis region of the spectrum, leading to enhanced reactivity [31]. Moreover, by decorating iron oxide with Au nanoparticles, we discovered that the nanocomposite material $Fe_2O_3 - Au$ efficiently transduces heat from light through plasmonic absorbance more efficiently than Au alone. This phenomenon was exploited to demonstrate the photothermal catalytic reduction of 4-nitrophenol [26]. Iron-based nanomaterials were successfully employed for removal of organic compounds, heavy metals, and radioactive contaminants [32].

Three different metallic nanoparticles, namely, Cu, Sn, and Ag were produced onto Fe/FexOy materials by straightforward wet chemical reduction methods. Basically, metal salts were reduced in water, in air with a reducing agent to yield spherical nanoparticles onto the porous iron substrates [11]. Nanoparticle production follows the typical four steps: [12] (a) nucleation step, (b) growth, (c) ripening, and (d) rapid consumption of residual precursors [33]. The experimental parameters used during the synthesis procedure, such as reduction rates of the metal precursors, temperature, reductant-to-precursor ratio, ligands, and strength of reductant, all have an effect on the final product. More exactly, the specific geometry, size, and crystallinity can be manipulated by the selection of these components. Therefore, minute changes to the amount, order, or timing significantly impact the nanoparticle's optical and physical properties.

Decoration of Cu, Sn, and Ag nanoparticles onto Fe/FexOy was monitored by scanning electron microscopy (SEM) (Fig. 2). Despite the diverse non-uniform surface condition, SEM images show that all nanoparticles are distributed over the entire Fe surface, including inside the pores of these structures. Noticeable differences in nanoparticles size and morphologies were recorded. In the case of Cu and Sn, the CTAB procedure produces well-defined nanostructures monodispersed in size with diameters of approximatively 100 nm on the surface of iron structures. In comparison, the citrate approach generates irregular nanostructures with a fair set of aggregates and well-dispersed structures on the support.

In the case of the silver nanostructures, however, the citrate approach generated better defined nanostructures with diameters in the range of 250 and 150 nm for Cu and Sn, respectively. A very limited number of aggregates were produced during the citrate approach. The presence of the metal nanoparticle on the iron structures often leads to a more efficient flow of electrons from the sorbent to the environment leading to faster processes. Additionally, addition of a secondary metal on the support increases the lifetime and reusability of the sorbent. Additionally, a limited amount of nanomaterial is needed to enhance the sorbent's efficiency. These benefits coupled with one's ability to produce nanomaterials via solution chemistries reduce material's cost and limit the amount of waste produced.

Particle-size distribution determined by measuring individual particles and clustered nanoparticles using scanning electron micrographs is displayed in Table 1. Moreover, the elemental composition of the nanocomposite structures was evaluated by the energy-dispersive X-ray analysis (EDS). These results confirm the presence of elements of interest, Cu, Sn, Ag, and Fe, as described in Table 1.While EDS is a semi-quantitative analysis, the data shows that the amount of nanoparticle loading varies based on the procedure used and elemental composition. In the case of Cu and Sn, the loading capacity doubled when the CTAB synthesis approach was used. That could be related to CTAB's ability to form micelle and/or promote electrostatic interactions with the iron support. It is certainly possible that the porous iron surface is oxidized with reticent iron oxide/hydroxide ions on the surface due to deprotonated Fe-OH surfaces. This is in agreement with our previous studies in which iron oxide nanoparticles produced by burning iron metal had a negative surface charge of -11 mV. In the case of silver nanoparticles, however, the loading capacity is five times larger when the citrate approach is used when compared with CTAB approach. It is believed that the nanoparticle growth mechanism is different in each case as it could relate to the CTAB tail contribution to the free energy of formation of a bilayer on different metallic nanoparticles [11]. The magnetic properties of the resulting nanocomposite iron support remained intact upon nanomaterial decoration. This key advantage can be strategically used for recovery via the use of a magnet. This is especially beneficial when there is a need to collect and dispose "payloads" at desired location. A representative EDS and EDS mapping of two different materials

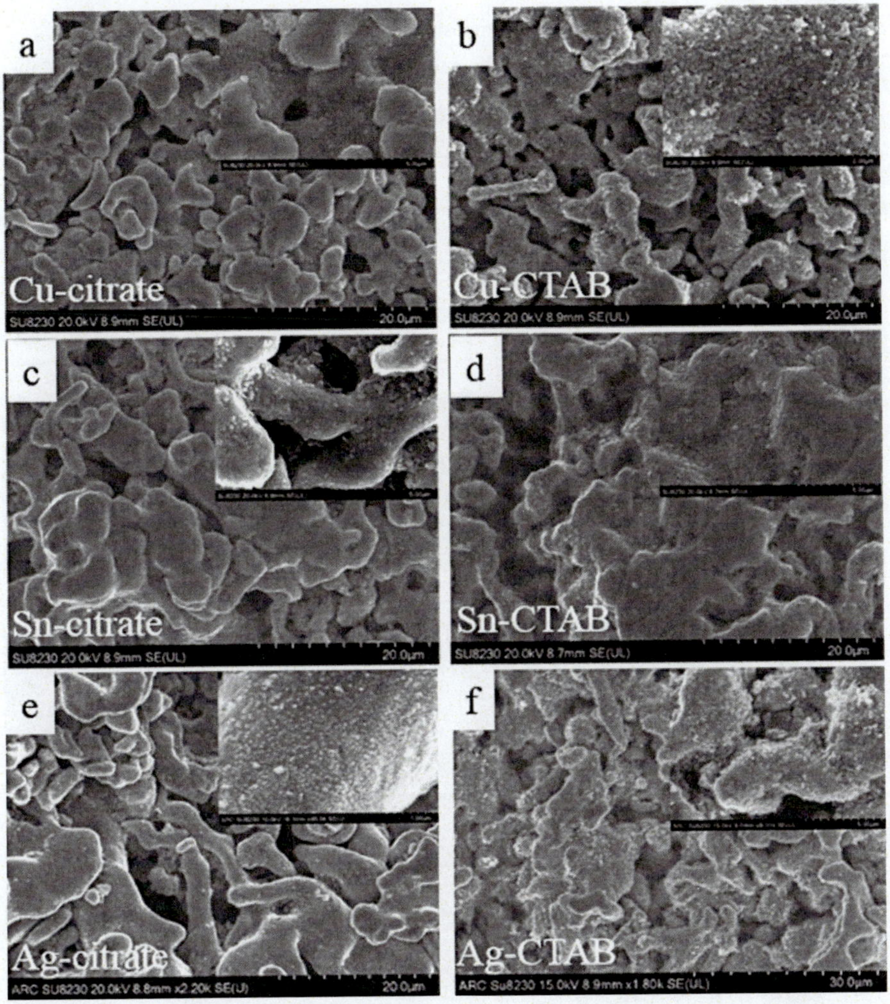

Fig. 2 Electron micrographs of nanomaterials prepared via citrate (left) and CTAB (right) approach: **a**, **b** Cu-Fe/FexOy nanocomposite; **c**, **d** Sn-Fe/FexOy nanocomposite; **e**, **f** Ag-Fe/FexOy nanocomposites

confirmed the presence of the nanomaterials on the entire surface of the iron support (Fig. 3).

Ligand Decoration of Fe/FexOy Structures

The use of different ligands generates materials with different properties. The tunability offered by the ligand's surface charge, i.e. positive, negative, and ligand's

Table 1 Nanoparticles dimensions and compositional analysis

Nanoparticles synthesis procedures	Dimensions (nm)	EDS data ratio (Metal/Fe %)
Cu-citrate	250	0.5
Cu-CTAB	100	1
Sn-citrate	150	0.2
Sn-CTAB	100	0.5
Ag-citrate	30	2
Ag-CTAB	150	0.4

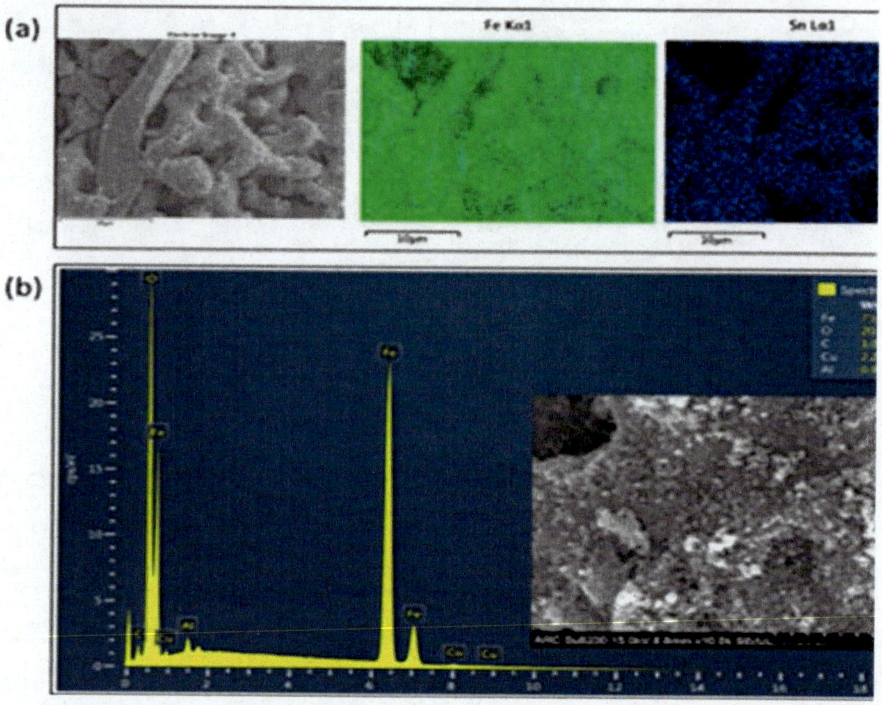

Fig. 3 **a** Representative EDS mapping of Sn-Fe/FexOy nanocomposite showing complete deposition, and **b** EDS and SEM image of Cu-Fe/FexOy nanocomposites

size, or chelation, can be further explored for sequestration of analytes of interest [34]. For example, different surface reactivities, i.e. ligands, but identical composition nanomaterial can be used to target various moieties through tailored electrostatic interactions. The stability and chemistry of the surface ligands can be used to selectively tune analytes' partitioning and permeability.

Two different synthesis approaches were employed in this study to generate bimetallic composite materials: a citrate approach and a CTAB approach. Citrate and

CTAB (Fig. 4) are the most common surfactants used for preparation of nanoparticles. In the case of the citrate approach, sodium citrate serves as both a reducing and a caping agent. In the case of CTAB, a strong reducing agent, sodium borohydride was used. There are major differences between these two approaches. In the citrate synthesis approach, the nanoparticles' surface is negative due to carboxyl groups' capping reagent which renders a negative effective surface charge (–37 mV). This opens the door to metal ion complexation and/or nonspecific electrostatic interactions with the negatively charged particle surface [35]. In contrast, the CTAB synthesis approach generates nanoparticles with a positive surface charge, typically + 30 mV. CTAB has a polar trimethylammonium group at one end making it soluble in aqueous systems and a 16-carbon cetyl tail that has a hydrophobic characteristics. The hydrophobic tail of CTAB interdigitates creating a "zipper-like" bilayer on the nanoparticle's surface. Therefore, the cationic head groups of CTAB are exposed to the environment and convey a positive surface charge of the nanoparticles. The hydrophobic layer can be used to partition water-insoluble moieties from the aqueous environments. CTAB is a micelle directing agent used in the preparation of shape selective metallic nanoparticle, i.e. rods, cubes, and triangles.

By employing these experimental conditions, one could not only preserve the structural integrity of Fe/Fe_xO_y, which is critical in this study but also stimulate specific surface interactions. Surface ligand's molecular structure, length, charge, etc. have a profound effect on material's reactivity, behavior, and stability [36]. Diffusion, interfacial solvent, counterions, and local clustering are just a few examples of parameters that can be used to modulate nanomaterial's surface heterogeneity, electrostatics, and preferential interactions [37]. Solvated ions, interfacial water, and ion distribution have an impact on the biding affinity of nanoparticles to various moieties [38].

Conclusions

We demonstrated production of nanocomposite materials, Cu, Sn, and Ag on Fe/FexOy materials through two different reduction approaches, i.e. citrate or cetyltrimethylammonium bromide (CTAB) approach. These procedures lead to the production of materials with different surface properties which can be further explored for selective sequestration of analytes of interest. Depending on the procedure employed, the nanoparticle's amount on Fe/FexOy varied. For example, metal atom concentration increases as follows: Ag < Sn < Cu when the CTAB approach was used. However, in the citrate approach, the highest metal load was silver with a trend of Sn < Cu < Ag. The preparation procedure is straightforward, cost-effective, and materials can be deployed in the field. The Fe/FexOy material retains its integrity and magnetic properties upon second nanometal manufacture. Therefore, these materials can be easily manipulated and placed as strategic locations through the use of a magnet.

Acknowledgements This work was supported by the Laboratory Directed Research and Development (LDRD) program within the Savannah River National Laboratory (SRNL). This work was produced by Battelle Savannah River Alliance, LLC under contract no. 89303321CEM000080 with the U.S. Department of Energy. Publisher acknowledges the U.S. Government license to provide public access under the DOE Public Access Plan (http://energy.gov/downloads/doe-public-access-plan).

References

1. Hunyadi Murph SE, Murph MA (2023) Nuclear fusion: the promise of endless energy. Phys Sci Rev 8(10):3095–3118. https://doi.org/10.1515/psr-2021-0069
2. Chew D et al (2016) SRR-LWP-2009–00001, rev.20, Savannah River Site
3. Gerdes K et al (2009) SRS salt waste processing facility technology readiness assessment
4. Hunyadi Murph SE et al (2017) Nanoparticle treated stainless steel filters for metal vapor sequestration. JOM 69:162–172. https://doi.org/10.1007/s11837-016-2206-5
5. Larsen G et al, Radiation detectors employing contemporaneous detection and decontamination, US20200395140A1. https://patents.google.com/patent/US20200082954A1/en
6. Hunyadi Murph SE et al, Detection and mitigation of radionuclides in the environment: toward a clean ecosystem.TMS 2023 152nd annual meeting & exhibition supplemental proceedings, pp. 715–724. https://doi.org/10.1007/978-3-031-22524-6_64
7. Fondeour F et al (2020) Methods and materials for determination of distribution coefficients for separation materials, 10:598–599. United States Patent: 10598599 (uspto.gov). https://patents.google.com/patent/US10598599B2/en
8. Li D et al (2019) Porous iron material for TcO4- and ReO4- sequestration from groundwater under ambient oxic conditions. J Hazard Mater 374:177–185. https://doi.org/10.1016/j.jhazmat.2019.04.030
9. Mortazavian S et al (2022) Biochar nanocomposite as an inexpensive and highly efficient carbonaceous adsorbent for hexavalent chromium removal. Materials 15(17):6055. https://doi.org/10.3390/ma15176055
10. Hunyadi Murph SE, Schyck S, Lawrence K (2022) Engineered nano-antenna susceptor as efficient platforms for efficient uptake and release of analytes. In: Srivatsan TS, Rohatgi PK, Hunyadi Murph S (eds) Metal-matrix composites. The minerals, metals & materials series. Springer, Cham. https://doi.org/10.1007/978-3-030-92567-3_22
11. Hunyadi Murph SE, Larsen G, Coopersmith K (2017) Anisotropic and shape-selective nanomaterials: structure-property relationships. Springer Nature, pp 1–470. https://doi.org/10.1007/978-3-319-59662-4
12. Hunyadi Murph SE et al (2013) Patchy silica-coated silver nanowires as SERS substrates. J. Nanopart Res 15(6):1607. https://doi.org/10.1007/s11051-013-1607-4
13. Hunyadi Murph SE, Gaulden P (2022) Novel nanophotocatalysts for detection and remediation of contaminated ecosystems. In: Srivatsan TS, Rohatgi PK, Hunyadi Murph S (eds) Metal-matrix composites. The minerals, metals & materials series. Springer, Cham, pp 335–349. https://doi.org/10.1007/978-3-030-92567-3_21
14. Tao C et al (2007) Surface morphology and step fluctuations on silver nanowires. Surf Sci 601:4939. https://doi.org/10.1016/j.susc.2007.08.023
15. Hunyadi Murph SE et al (2011) Synthesis, functionalization, characterization and application of controlled shape nanoparticles in energy production, fluorine-related nanoscience with energy applications. ACS Symposium Series, vol 1064, Chapter 8, pp 127–163. https://doi.org/10.1021/bk-2011-1064.ch008
16. Hunyadi Murph SE et al (2012) Metallic and hybrid nanostructures: fundamentals and applications, in applications of nanomaterials. In: Govil JN (ed) Nanomaterials and nanostructures. series Studium Press LLC, USA

17. Hunyadi Murph SE, Majidi V (2021) Systems and methods for manufacturing nano-scale materials US Patent US20210380405A1. https://patents.google.com/patent/US20210380405A1/en
18. Greenlee LF et al (2012) Kinetics of zero valent iron nanoparticle oxidation in oxygenated water. Environ Sci Technol 46:12913–12920
19. Lien HL, Zhang WX (2007) Nanoscale Pd/Fe bimetallic particles: catalytic effects of palladium on hydrodechlorination. Appl Catal B – Environ 77:110–116
20. Wang CB, Zhang WX (1997) Synthesizing nanoscale iron particles for rapid and complete dechlorination of TCE and PCBs. Environ Sci Technol 31:2154–2156
21. Crane RA, Pullin H, Macfarlane J, Silion M, Popescu IC, Andersen M, Calen V, Scott TB (2015) Field application of iron and iron-nickel nanoparticles for the ex situ remediation of a uranium-bearing mine water effluent. J Environ Eng 141(8)
22. Hu CY, Lo SL, Liou YH, Hsu YW, Shih KM, Lin CJ (2010) Hexavalent chromium removal from near natural water by copper-iron bimetallic particles. Water Res 44(10):3101–3108
23. Zhou T, Li YZ, Lim TT (2010) Catalytic hydrodechlorination of chlorophenols by Pd/Fe nanoparticles: comparisons with other bimetallic systems, kinetics and mechanism. Sep Purif Technol 76(2):206–214
24. Hunyadi Murph SE (2011) One-dimensional plasmonic nano-photocatalysts: synthesis, characterization and photocatalytic activity. In: Tachibana Y (ed) Solar hydrogen and nanotechnology VI. Proceedings of SPIE. vol 8109, 81090T, pp 1–11. https://doi.org/10.1117/12.893029
25. Hunyadi Murph SE, Murphy CJ (2013) Patchy silica-coated silver nanowires as SERS substrates. J Nanoparticle Res 15(6):1607. https://doi.org/10.1007/s11051-013-1607-4
26. Hunyadi Murph SE, Larsen G, Lascola R (2016) Multifunctional hybrid Fe2O3-Au nanoparticles for efficient plasmonic heating. J Visual Exp (JOVE) 108:e53598. https://doi.org/10.3791/53598
27. Hunyadi Murph SE, Lawrence K, Sessions H, Brown M, Larsen G (2020) Controlled release of hydrogen isotopes from hydride-magnetic nanomaterials. ACS Appl Mater Interfaces 12(8):9478–9488. https://doi.org/10.1021/acsami.0c00887
28. Hunyadi Murph SE et al (2021) Efficient thermal processes using alternating electromagnetic field for methodical and selective release of hydrogen isotopes. Energy Fuels 35:3438–3448. https://doi.org/10.1021/acs.energyfuels.0c03704
29. Hunyadi Murph SE et al, Nanocomposite materials for accelerating decarbonization. TMS 2023 152nd annual meeting & exhibition supplemental proceedings, pp. 748–757. https://doi.org/10.1007/978-3-031-22524-6_68
30. Hunyadi Murph SE (2020) Shape-selective mesoscale nanoarchitectures: preparation and photocatalytic performance. Catalysts 10:532. https://doi.org/10.1007/978-3-319-59662-4
31. Srivatsan TS, Harrigan WC, Hunyadi Murph SE (2021) Metal-matrix composites: advances in analysis, measurement, and observations, p 281. https://doi.org/10.1007/978-3-030-65249-4
32. Srivatsan TS, Rohatgi PK, Hunyadi Murph SE (2022) Metal-matrix composites: advances in processing, characterization, performance and analysis. Springer Nature, pp 1–385. https://doi.org/10.1007/978-3-030-92567-3
33. Larsen G et al (2016) Multifunctional Fe2O3-Au nanoparticles with different shapes: enhanced catalysis, photothermal effects, and magnetic recyclability. J Phys Chem C 120:15162–15172. https://doi.org/10.1021/acs.jpcc.6b03733
34. Gankanda A, Rentz NS, Greenlee LF (2019) Greenlee influence of ligand size and chelation strength on zerovalent iron nanoparticle adsorption and oxidation behavior in the presence of water vapor and liquid water. J Phys Chem C 123(4):2474–2487
35. Hunyadi Murph SE (2012) Manganese–doped gold nanoparticles as positive contrast agents for magnetic resonance imaging (MRI). J Nanopart Res 14:658–659. https://doi.org/10.1007/s11051-011-0658-7
36. Cui AY, Cui Q (2021) Modulation of nanoparticle diffusion by surface ligand length and charge: analysis with molecular dynamics simulations. J Phys Chem B 125(17):4555–4565. https://doi.org/10.1021/acs.jpcb.1c01189

37. Liang D et al (2020) Ligand length and surface curvature modulate nanoparticle surface heterogeneity and electrostatics. J Phys Chem C 124:24513–24525. https://doi.org/10.1021/acs.jpcc.0c08387
38. Liang, D at al (2020) Interfacial water and ion distribution determine ζ potential and binding affinity of nanoparticles to biomolecules. Nanoscale 12(35):18106–18123. https://doi.org/10.1039/D0NR03792C
39. Hunyadi Murph SE, Larsen G, Coopersmith K (2017) Anisotropic and shape-selective nanomaterials: structure-property relationships. Springer Nature. https://doi.org/10.1007/978-3-319-59662-4

Photocatalytic Nitrate Destruction Studies in Complex Environments

Simona E. Hunyadi Murph

Abstract One method for denitration of nitric acid used in nuclear facilities is to use formic acid as a reductant. The major problem with formic acid denitration is an induction period of varying duration that may result in excessive accumulation of formic acid at the reaction onset. This accumulation poses an off-gas process control issue. In this paper, we will describe the use of titania-based photocatalysts for the treatment of nitric acid and nitrate wastes. We find that the photocatalytic process is a simple and straightforward method to completely destroy nitrate ions at room temperature without any initiation period.

Keywords Photocatalysts · Nanomaterials · Nitrate · Nanocomposite

Introduction

The anthropogenic pollutants, such as pesticide, nuclear materials, organic contaminants, oils, fertilizers, and heavy metals, are a global concern as they pose a hazard to human health and our ecosystems [1–6]. Pollutants are found in wastewater, air, and land environments. Therefore, a wide range of scientific and engineering innovations are needed to efficiently meet the regulatory requirements of clean soil, water, and atmospheric environments.

Management of nuclear materials in a safe, secure, and effective way is a multifaceted and operationally complex process [7]. It is based on intricate sequential multistep processes conducted in specialized nuclear facilities. For example, H-Canyon is a production-scale specialized facility located at Savannah River Site in which radiologically shielded chemical separations are being conducted [8]. While at its inception, in the 1950s, the focus of this facility was on the production of

S. E. Hunyadi Murph (✉)
Environmental and Legacy Management Directorate, Savannah River National Laboratory, Aiken, SC, USA
e-mail: Simona.Murph@srnl.doe.gov

Department of Physics and Astronomy, University of Georgia, Athens, GA, USA

B. Wisner et al. (eds.), *Composite Materials*, The Minerals, Metals & Materials Series,
https://doi.org/10.1007/978-3-031-50180-7_6

radioactive materials for use in nuclear weapons, currently its mission has shifted to nonproliferation and environmental activities. The main focus today is on clean-up efforts and regulatory compliance of environmental legacy radioactive waste, contaminated land and assets, that resulted from decades of nuclear weapons production and nuclear energy research. A number of efficient and deployable technologies and strategies are being explored and implemented at various contaminated sites to reduce cost and accelerate clean-up efforts [1–7, 9].

Nitrate moieties represent a significant fraction of the waste requiring treatment [10]. Chemical processes are often deployed to remove nitric acid from waste-contaminated plutonium and uranyl solutions [11, 12]. Typically, uranium and plutonium processing starts by dissolution of the uranium fuel rods encased in aluminum using nitric acid that is heated to a boil. The resulting acidic solution contains a massive number of analytes and chemical compounds, such as aluminum, uranium, plutonium, neptunium, mercury, fission products, and other impurities that necessitate further processing and handling. Therefore, treatment and handling of this multifaceted environment, i.e., nuclear materials, low/high pH environments, multitude of contaminants and by-products, etc., is based on comprehensive chemical separation stages and flowsheet processes. For example, the low pH environments of the waste stream generated during the dissolution process, in which radioactive materials are present, represent a serious problem in the nuclear fuel reprocessing and it affects the entire process. Caustic neutralization being a potential neutralization solution adds significant unwanted substances (cations and water) to the process. It also significantly increases the contaminated waste volume to be handled. To reduce the volume of high-level radioactive waste in storage, the liquid waste is routed through evaporators. A series of multi-operational physical/chemical treatment steps which include pH adjustment, filtration, reverse osmosis, etc. are also performed. These processes are operationally complex and energy intensive.

Nitrates/NO_X abatement represents a significant cost and risk for thermal waste treatment and long-term groundwater treatment processes [13]. Over the years, a number of reducing analytes were investigated as denitration agents, including formaldehyde, dextrose, sucrose, and formic acid [14, 15]. Each process, while operational, suffers from several drawbacks. For example, when formaldehyde is used, foaming and self-polymerization occurs requiring the addition of large volumes of water. Due to its rapid reactivity, formaldehyde may also cause off-gas process control problems. Dextrose and sucrose, while non-hazardous, require excessive reaction times leading to crystallization and radionuclide precipitations resulting in pipe line blockage and clogging. Formic acid, as an alternative solution, was also employed as a reductant for denitration of nitric acid. However, there are some disadvantages of the formic acid denitration process. A large amount of gas and heat are evolved rapidly if an excess of formic acid accumulates in concentrated nitric acid and reacts. Moreover, the process is not normally capable of destroying all of the nitrate ions and generally is inefficient for HNO_3 concentrations less than 0.5 M. Consequently, substantial concentrations of nitrate could still be present after formic acid treatment. One of the major problems with denitration using formic acid is an induction period of varying duration that may result in excessive accumulation of NO_x gases, which is an air pollutant, and N_2O, which is a greenhouse gas. These accumulations pose an off-gas process control issue [16].

Given the considerations above and emerging space constraints at nuclear facilities, a new denitration process is necessary to reduce the volume of nitric acid and nitrate-bearing waste generated from H-Canyon processes. Of particular importance is that the denitration process should have a short initiation period, have the capability of destroying most, if not all, of the nitrate, and not produce any by-products that will cause processing problems in downstream storage tanks and processing facilities. Further, an ideal solution must be safe, simple, cost-effective, and capable of being scalable to meet the demands of production-scale facilities. Photocatalysis is a promising denitration process that could meet all of these demands. The efficiency of these processes can be improved when using nano- or macro-scale materials [16–19]. However, this process has not been investigated for highly acidic nitrate-bearing wastes or complex environments. One of the most efficient photocatalytic materials studied is titania (TiO_2). This is due to its powerful oxidation capability, superior charge transport, and corrosion resistance. Additionally, TiO_2 is inexpensive and is abundant. Its limitations, however, are related to its bandgap (3.2 eV) which makes it responsive to UV illumination only. Bandgap engineering has been explored by doping and co-doping of titania with metals and non-metals [1, 20–24]. Noble metals, such as gold, increase titania's efficiency by expanding its response in the Vis region of the spectrum due to its plasmonic resonance [24–28]. In here, we explore the use of titania-based nanocatalysts, such as composite titania-gold nanoparticles and bare titania nanoparticles, for the treatment of nitrate wastes.

Experimental Details

Materials and Instrumentation

Chloroauric acid trihydrate, sodium nitrate ($NaNO_3$, 99%), and Degussa P-25 titania (TiO_2) anatase 25 nm spheres were purchased from Sigma Aldrich. All chemicals were used as received. Methanol was used as a hole scavenger. The hole scavengers influence the photocatalytic reaction by participating in the reduction of intermediates. Methanol donates its electrons to the photocatalyst during the illumination process (when holes are created), and therefore regenerates the catalyst. Deionized pure water was used throughout (18.3 MΩ cm). All glasswares were cleaned with aqua regia, and thoroughly rinsed with deionized water prior to use. 25 nm TiO_2 particles were used as photocatalyst. The nanomaterials are characterized using a scanning electron microscope (Hitachi SU8320). UV–vis spectroscopy was performed with Varian model Cary 500 Scan UV–Vis-NIR spectrophotometer. Light scattering and ξ-potential measurements were performed with a Brookhaven Zeta PALS instrument. Low-resolution transmission electron microscope (TEM) images were acquired with a 200 kV Hitachi H-8000 instrument. SEM measurements of both powdered and

3D-printed materials were performed using a Hitachi 8230 coupled with an Oxford Instruments X-Max 150 Silicon Drift Detector (EDS).

Synthesis of Plasmonic TiO_2–Au Nanocomposite

Degussa P-25 TiO_2 particles (25 nm) were used as support for gold (Au) metal deposition. Gold loading on titania nanoparticles was achieved via photo-deposition under UV light. Bare TiO_2 particles were suspended in an aqueous solution (1 cm^3) and were photo-irradiated in the presence of Au^{3+} ions for 5, 30, and 60 min, respectively. The photo-irradiation resulted in deposition of Au onto TiO_2. The preparation procedure was conducted at two different pHs: 3 and 10. The pH of the suspension was adjusted to the desired value using sodium hydroxide solution. A UV light (365 nm LED) radiation source was used.

Photocatalytic Activity

The photocatalytic activity of bare titania and as-synthesized TiO_2–Au nanomaterials was evaluated in the photodegradation of nitrate (0.1 M) in water. The test was carried out by using a UV light (365 nm LED) radiation source. Bare TiO_2 or metal-loaded TiO_2 aqueous particles were suspended in a solution (1 cm^3) containing $NaNO_3$ 0.1 M in a plastic reactor. The decomposition of nitrate of 0.1 M was conducted in a 1 ml cylindrical quartz photo-reactor under continuously stirring. A blank solution was prepared by the same method without nanophotocatalyst. At regular intervals (30 min), the nitrate-nanophotocatalyst solution was sampled and its concentration measured by its absorbance at 305 nm by a UV–visible spectrophotometer. Prior to the irradiation, each sample was kept in dark for 30 min, to ensure equilibration.

Results and Discussion

The use of photocatalytic materials for degradation of organic and inorganic pollutants found in wastewater is a suitable green and economical strategy that can be deployed at contaminated sites [1, 18, 19, 28, 29]. The photocatalytic process is straightforward and involves the use of a photo-responsive material and solar energy. The degradation process occurs when hydroxyl radicals are generated upon electromagnetic exposure, i.e., UV–vis illumination. Typically, heterogeneous photocatalysis involves the excitation of light of photoinduced reactions which occur at the surface of a semiconducting catalyst (e.g., TiO_2, ZnO, Fe_xO_y, etc.) or via active species generated on the surface and transferred to solution. These photogenerated charge carriers enable reduction and oxidation reactions in the surrounding environment. Often, a hole scavenger is used to improve the photodegradation process. The hole scavenger is a sacrificial chemical which is oxidized in the photocatalytic process and thus removes the hole from the TiO_2 nanostructure. This enhances photocatalytic processes by preventing TiO_2 electron–hole recombination and therefore increases the lifetime of the other charge carriers. Ultimately, a hole scavenger (i.e., electron donor) stabilizes the nanocatalyst against photooxidation and decreases the rate of recombination allowing more electrons to become available for reduction reactions. The degradation of pollutants is easily monitored by UV–vis spectroscopy and is often used as a simple model to evaluate the efficacy of potential photocatalytic materials. The use of titania-based photocatalysts for the selective reduction of nitrates to nitrogen gas (N_2) has been used in groundwater remediation efforts to clean up nitrates from high agricultural areas [30–32].

Nanophotocatalyst Preparation and Characterization

An efficient photocatalysis process requires a large fraction of the photogenerated electrons and holes to separate. Once separated, these species must transport to the redox reaction sites, i.e., surface, before recombination. Therefore, nanoscale titania materials are advantageous when compared with bulk size titania. This is due to enhanced mobility and low probability of electron–hole recombination of the photoinduced electron–hole generated. Nanomaterials have a higher reactive surface area than their bulk counterparts, which allows for reduced catalyst load requirements, reduced overall material cost, and reduced waste generation, all without compromising performance. Titania can be found in three different crystal phases, namely, anatase, rutile, and brookite. Most studies suggest that the anatase phase of TiO_2 outperforms the other phases. However, other investigations show that a mixture of different crystalline phases could provide even better photocatalytic performance. For example, a combination of anatase and rutile TiO_2 at an appropriate ratio can outperform either pure anatase or pure rutile [33]. A mixture of 70% anatase and 30% rutile, known as Degussa P-25 TiO_2, shows even better photodegradation efficiency compared to other forms. Therefore, Degussa P-25 Titania nanospheres were used in our studies.

Particle-size distribution, determined by measuring individual particles on scanning electron micrographs, was found to be 37 ± 2 nm ($n = 50$) for the Degussa P-25 TiO_2 nanoparticles (NPs). This is in disagreement with the size reported by the vendor, 25 nm. Smaller size nanomaterials are ideal for catalytic applications as they possess much higher reactive surface areas than larger colloids. It is well known that the smaller the particle the larger the fraction of atoms at the surface, and the higher the average binding energy per atom. Nanomaterials' surface charge and hydrodynamic radius in solution were measured by dynamic light scattering (DLS).

A multimodal size distribution data analysis collected on titania NPs solution (sonicated for 10 min in aqueous solution, at a pH = 3) shows an average distribution of hydrodynamic titania nanoclusters with diameters of 432 nm ± 56 nm.

The multimodal distributions showing multiple peaks demonstrate that that the sample is polydisperse in nature containing nanocluster particulates of various sizes: 90 nm, ~500 nm, and ~1000 nm. This is not surprising, since the titania nanoparticulates are "surfactantless" and are susceptible to aggregation. Solution pH affects the dispersion hydrodynamic diameter by changing the particle surface charge. The pH at which the surface of titania is neutral is point of zero charge or isoelectric point. The isoelectric point (IEP) for TiO_2 (P-25) is approximately 6.2 [34]. Near isoelectric point, significant agglomeration takes place; large flocs were observed, as the particle surface charge is close to zero and attractive van der Waals forces are dominant. When the pH is different from IEP for titania, the absolute value of zeta potential becomes higher and the hydrodynamic size becomes smaller.

The pH value of the solution plays a significant role in the interactions between the surface of TiO_2 and the instinctive electron property of the degrading substance. At pH values that are less than the pH of zero point charge, the major species on the surface of catalysts is in the form of $Ti–OH_2^+$. When the pH is greater than pH of zero point charge, the negative form of $Ti–O^-$ is present. Therefore, the pH value of the solution plays a significant role in the interactions between the surface of TiO_2 and the instinctive electron property of the degrading substance. Phase Analysis Light Scattering (PALS) techniques were employed to evaluate the effective surface charge of the titania NPs as function of pH. As described above, at a low pH where $Ti–OH_2^+$ species are expected in the aqueous solution, a positive surface charge specifically $\zeta =$ +17 mV was recorded. This is consistent with previously reported results of surface charging of metal oxides in electrolytic solutions, which is generally attributed to the amphoteric character of surface hydroxyl groups [35–37].

UV–vis spectroscopy was used to characterize the titania nanoparticles (Fig. 1). In DI water, the UV–vis spectra contain a characteristic peak at 330 nm that corresponds to the bandgap transition of TiO_2 semiconductor (~3.2 eV). As expected, TiO_2 has no absorption band in visible range and shows the characteristic spectrum with its fundamental absorption of Ti–O bond in ultraviolet light range. Gold nanospherical particles were grown onto TiO_2 nanospheres by photocatalytic reduction of $AuCl_4^-$ ions after direct UV illumination. The pH solution was maintained at 3. Au NPs formation was monitored at various intervals, 5, 30, and 60 min via UV–vis spectroscopy. A quick visual analysis of the solution revealed the formation of gold nanostructures as a purple solution emerged. A peak corresponding to the plasmon resonance for Au NPs at 590 nm was collected via UV–vis spectroscopy analysis. This suggests the formation of large Au nanostructures and/or aggregates. SEM analysis (Fig. 1) confirms the formation of Au nanospherical deposits with variable diameters of 50 nm and formation of aggregates and elongated nanostructures (>100 nm).

Fig. 1 **a** SEM images of titania nanoparticles; **b** UV–vis spectra of TiO_2-based nanoparticles; and **c–d** SEM images of titania-gold nanoparticles (backscattered)

Photodegradation Activity

In order to study the effect of TiO_2 and TiO_2–Au nanomaterials on the photocatalytic activities, nitrate photodegradation experiments were performed under UV illumination. TiO_2 and TiO_2–Au nanoparticles were first evaluated for nitrate degradation under UV irradiation. The photocatalytic degradation of nitrate ions in solution was first tested using $NaNO_3$ as the analyte source. $NaNO_3$ was selected not only because it is a relevant source of nitrate, but also because it is directly relevant to waste accumulation from H-Canyon waste streams. $NaNO_3$ is a product of the neutralization of any unreacted nitric acid, $HNO_3 + NaOH \rightarrow NaNO_3 + H_2O$, and it cannot be reduced by traditional reductants (e.g., formic acid) and is the most difficult nitrate source to remove. The titania photocatalyst can reduce nitrate ions when its electrons are excited by the UV illumination to a free state in the conduction band. Once freed, these conduction band electrons can be transferred to nitrate ions adsorbed to the photocatalyst surface or in the electric double layer. Concurrently, the excitation of the electron creates a hole in the valence band of the photocatalyst (i.e., electron/hole pair). To maintain electroneutrality, the hole must be filled with an electron either through the oxidation of compounds (i.e., electron donor) or from recombination of the electron/hole pair. Recombination is not desired, as this results in a loss of the photogenerated electron/hole pair, causing a decrease in photocatalytic activity.

Adding a hole scavenger (HS) (i.e., electron donor) decreases the rate of recombination, which causes more electrons to become available for reduction reactions. Varying the amount and type of hole scavenger determines the photoreduction pathway and the speed of the process. A candidate hole scavenger, methanol, was investigated in this study. Using methanol as a hole scavenger is beneficial as the reaction products are gases, i.e., CO_2 and H_2 [38]. When methanol and nitrate were mixed under dark or irradiated conditions, no nitrate reduction occurred without a photocatalyst. Therefore, methanol does not act as a direct reductant under these experimental conditions; rather, it acts as a hole scavenger. Another way to decrease the recombination rate is to add a metal, i.e., Au, co-catalyst to the surface of the photocatalyst. The metal can act as an electron trap or sink for conduction band electrons, thus increasing the time available for the hole to be filled and the electrons to complete the interfacial transfer. Doping of titania Au nanoparticles has also been investigated as a means for improving the photocatalytic properties of titania and extending the response of the material into the visible region [39]. Au nanoparticles are of special interest in this area because they efficiently absorb and scatter light in the visible region of the electromagnetic (EM) spectrum. This is due to localized surface plasmon resonance (LSPR). In order to elucidate the photocatalytic behavior of the composite TiO_2–Au, both UV and visible light photodegradation experiments were performed. Methanol was used in various increments and at various times to increase the ratio of methanol: nitrate ions. A series of photodegradation experiments were conducted and are summarized in Fig. 2. All experiments showed a continued photodegradation of nitrate ions as measured by UV–vis spectroscopy.

These results shows that the most efficient photocatalytic experiment was when the composite TiO_2–Au was used under UV–vis illumination. Results demonstrate that the photodegradation process is most efficient, ~80–90% of 0.1 M nitrate photodegradation after 4–5 h of exposure, when a hybrid nanomaterial, TiO_2–Au, and a hole scavenger are used. The presence of the hole scavenger positively influences the

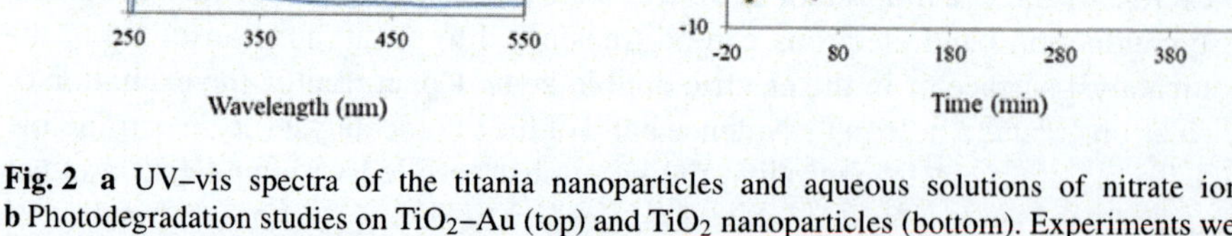

Fig. 2 **a** UV–vis spectra of the titania nanoparticles and aqueous solutions of nitrate ions. **b** Photodegradation studies on TiO_2–Au (top) and TiO_2 nanoparticles (bottom). Experiments were conducted in the presence and absence of hole scavenger. Typically, the presence of hole scavenger improves the photodegradation process

process. Additionally, increasing the amount of hole scavenger showed an increase in nitrate degradation.

Conclusions

The proof-of-concept studies conducted demonstrate the feasibility of this technology. Our studies showed ~80–90% of 0.1 M nitrate photodegradation after 4–5 h exposure in UV–Vis light with TiO_2–Au hybrid nanomaterials. It serves as a powerful predictor for potential application in nuclear facility as it is cost-effective and efficient. This technology reduces the variety and volumes of chemicals used during separations of nuclear waste through the use of nanoscale materials. The photocatalytic process does not require any external heating, does not exhibit an initiation period, and is capable of destroying refractory nitrates such as $NaNO_3$. Moreover, this process requires only the use of inexpensive photocatalysts combined with UV–vis light to make it feasible for large-scale treatment applications.

References

1. Murph SEH, Searles E (2021) Iron oxide–gold composite nanoparticles and nano-gap junctions for sensing applications using surface-enhanced Raman scattering. In: Srivatsan TS, Harrigan Jr. WC, Hunyadi Murph S (eds) Metal-matrix composites. The Minerals, Metals & Materials Series. Springer, Cham. https://doi.org/10.1007/978-3-030-65249-4_17
2. Li D, Seaman J et al (2019) Porous iron material for TcO_4^- and ReO_4^- sequestration from groundwater under ambient oxic conditions. J Hazard Mater 374:177–185. https://doi.org/10.1016/j.jhazmat.2019.04.030
3. Hunyadi Murph SE et al (2017) Nanoparticle treated stainless steel filters for metal vapor sequestration. JOM 69:162–172. https://doi.org/10.1007/s11837-016-2206-5
4. Hunyadi Murph SE et al (2023) Detection and mitigation of radionuclides in the environment: toward a clean ecosystem. In: TMS 2023 152nd annual meeting & exhibition supplemental proceedings, pp 715–724. https://doi.org/10.1007/978-3-031-22524-6_64
5. Mortazavian S et al (2022) Biochar nanocomposite as an inexpensive and highly efficient carbonaceous adsorbent for hexavalent chromium removal. Materials 15(17):6055. https://doi.org/10.3390/ma15176055
6. Hunyadi Murph SE, Lawrence K et al (2020) Controlled release of hydrogen isotopes from hydride-magnetic nanomaterials. ACS Appl Mater Interfaces 12(8):9478–9488. https://doi.org/10.1021/acsami.0c00887
7. Hunyadi Murph SE, Murph MA (2023) Nuclear fusion: the promise of endless energy. Phys Sci Rev 8(10):3095–3118. https://doi.org/10.1515/psr-2021-0069
8. https://savannahrivernuclearsolutions.com/programs/nuc_mtrls01.htm
9. Fendour F et al (2020) Methods and materials for determination of distribution coefficients for separation materials. US Patent 10,598,599. https://patents.google.com/patent/US10598599B2/en
10. Ando M et al (2021) A kinetic model for the autocatalytic behavior of nitric acid/formic acid mixtures to predict induction period. Process Saf Environ Prot 151:182–187
11. Hill D (2008) Nuclear energy for the future. Nature Mater 7:680–682. https://doi.org/10.1038/nmat2247

12. Glinskii ML et al (2020) Development of environmental monitoring in the vicinity of nuclear energy facilities. At Energy 127:166–173 (2020). https://doi.org/10.1007/s10512-020-00605-7
13. Pitt WW et al (1981) Biological reduction of nitrates in wastewaters from nuclear processing using a fluidized-bed bioreactor. Nucl Chem Waste Manage 2(1):57–70
14. Numata M et al (2006) Reduction of sodium nitrate liquid waste in nuclear reprocessing plans. In: Waste management conference, pp 1–9
15. National Academies of Sciences, Engineering, and Medicine (1996) Nuclear wastes: technologies for separations and transmutation. The National Academies Press, Washington, DC. https://doi.org/10.17226/4912
16. Ren G et al (2021) Recent advances of photocatalytic application in water treatment: a review. Nanomaterials (Basel). 11(7):1804. https://doi.org/10.3390/nano11071804. PMID: 34361190; PMCID: PMC8308214
17. Hunyadi Murph SE, Schyck S, Lawrence K (2022) Engineered nano-antenna susceptor as efficient platforms for efficient uptake and release of analytes. In: Srivatsan TS, Rohatgi PK, Hunyadi Murph S (eds) Metal-matrix composites. The minerals, metals & materials series. Springer, Cham. https://doi.org/10.1007/978-3-030-92567-3_22
18. Srivatsan T, Rohatgi PK, Hunyadi Murph SE (2022) Metal-matrix composites: advances in processing, characterization, performance and analysis. Springer Cham, vol XVI, p 385. https://doi.org/10.1007/978-3-030-92567-3
19. Hunyadi Murph SE et al (2021) Efficient thermal processes using alternating electromagnetic field for methodical and selective release of hydrogen isotopes. Energy Fuels 35:3438–3448. https://doi.org/10.1021/acs.energyfuels.0c03704
20. Hunyadi Murph SE et al (2020) Shape-selective mesoscale nanoarchitectures: preparation and photocatalytic performance. Catalysts 10:532. https://doi.org/10.3390/catal10050532
21. Hunyadi Murph SE et al (2011) Synthesis, functionalization, characterization and application of controlled shape nanoparticles in energy production, fluorine-related nanoscience with energy applications. ACS symposium series, vol 1064, chap 8, pp 127–163. https://doi.org/10.1021/bk-2011-1064.ch008
22. Hunyadi Murph SE et al (2012) Metallic and hybrid nanostructures: fundamentals and applications, in applications of nanomaterials. In: Govil JN (ed) Nanomaterials and nanostructures. series Studium Press LLC, USA
23. Hunyadi Murph SE et al (2023) Nanocomposite materials for accelerating decarbonization. In: TMS annual meeting & exhibition, pp 748–757. https://doi.org/10.1007/978-3-031-22524-6_68
24. Kun Y et al (2016) Fe_2O_3/TiO_2 core-shell nanorod array for visible light photocatalysis. Catal Today, special issue C1 Catalytic Chemistry, 51–58. https://doi.org/10.1016/j.cattod.2015.10.026
25. He Y, Basnet P et al (2013) Ag nanoparticle embedded TiO_2 composite nanorod arrays fabricated by oblique angle deposition: toward plasmonic photocatalysis. ACS Appl Mater Interfaces 5(22):11818–11827. https://doi.org/10.1021/am4035015
26. Hunyadi Murph SE (2011) One-dimensional plasmonic nano-photocatalysts: synthesis, characterization and photocatalytic activity, solar hydrogen and nanotechnology VI. In: Tachibana Y (ed) Proceedings of SPIE 2011, vol 8109, 81090T, pp 1–11
27. Hunyadi Murph SE et al (2013) Patchy silica-coated silver nanowires as SERS substrates. J Nanoparticle Res 15(6):1607. https://doi.org/10.1007/s11051-013-1607-4
28. Tao C et al (2007) Surface morphology and step fluctuations on silver nanowires. Surf Sci 601:4939. https://doi.org/10.1016/j.susc.2007.08.023
29. Hunyadi Murph SE, Gaulden P (2022) Novel nanophotocatalysts for detection and remediation of contaminated ecosystems. In: Srivatsan TS, Rohatgi PK, Hunyadi Murph S (eds) Metal-matrix composites. The Minerals, Metals & Materials Series. Springer, Cham., pp 335–349. https://doi.org/10.1007/978-3-030-92567-3_21
30. Hunyadi Murph SE, Larsen G, Lascola R (2016) Multifunctional hybrid Fe_2O_3–Au nanoparticles for efficient plasmonic heating. J Vis Exp (JOVE) 108:e53598. https://doi.org/10.3791/53598

31. Shand M et al (2013) Aqueous phase photocatalytic nitrate destruction using titania based materials: routes to enhanced performance and prospects for visible light activation. Catal Sci Technol 3:879
32. Doudrick K et al (2013) Photocatalytic nitrate reduction in water: managing the hole scavenger and reaction by-product selectivity. Appl Catal B Environ 136–137:40–47
33. Hurum DC et al (2003) Explaining the enhanced photocatalytic activity of Degussa P25 mixed-phase TiO_2 using EPR. J Phys Chem B 107(19):4545–4549
34. Hellstern HL et al (2016) Core–shell nanoparticles by silica coating of metal oxides in a dual-stage hydrothermal flow reactor. Chem Commun 52:3434
35. Venkatesan KA, Sati Sasidharan N, Wattal PK (1997) Specific adsorption of cesium on silica-titania. J Radioanal Nucl Chem 222:223–226 (1997). https://doi.org/10.1007/BF02034274
36. Borghi F et al (2013) Nanoscale roughness and morphology affect the isoelectric point of titania surfaces. PLoS One 8(7):e68655
37. Duval J et al (2002) Double layers at amphifunctionally electrified interfaces in the presence of electrolytes containing specifically adsorbing ions. J Electroanal Chem 532:337–435
38. Bowker M et al (2022) The role of metal nanoparticles in promoting photocatalysis by TiO_2. Top Curr Chem 380:17
39. Hunyadi Murph SE, Larsen G, Coopersmith K (2017) Anisotropic and shape-selective nanomaterials: structure-property relationships. Springer Nature. https://doi.org/10.1007/978-3-319-59662-4

Technical Route to Develop High-T_g Epoxy Composite That Is Water Degradable at Low Temperature

Lei Zhao, Jiaxiang Ren, Tim Dunne, and Peng Cheng

Abstract Degradable polymers and composites have found extensive applications in various industries but are currently limited to low-temperature usage due to their low glass transition temperature (typically <80 °C). High-T_g polymer composites, predominantly composed of epoxy resin, present challenges in decomposition, often necessitating high temperature, pressure, corrosive environments, and even supercritical fluids. The increasing adoption of fiber-enhanced composites for weight reduction and corrosion resistance further aggravates environmental concerns. This study addresses these challenges by introducing a pioneering solution in the form of a novel high-T_g (>130 °C) epoxy composite that exhibits water degradability below 100 °C. The material development process and degradation mechanism will be comprehensively elucidated. To the best of our knowledge, this marks the industry's first water-degradable high-T_g epoxy composite, unlocking a plethora of new applications across industries and providing an environmentally friendly, green composite material.

Keywords High T_g · Epoxy · Composite · Water degradable · Low temperature

Introduction

In the oil and gas industry, the demand for a degradable polymer composite material capable of functioning at high temperatures (typically >130 °C) and degrading in near-neutral aqueous environments at low temperatures (typically <100 °C) is essential for various equipment, including dissolvable hydraulic fracturing plugs [1–3]. Such materials can significantly enhance operation efficiency and reduce operational costs. Despite extensive investigations into degradable and dissolvable materials by academia and various industries, existing thermoplastic and thermoset systems have not proved suitable for this application due to their unsuitable combination of working and degradation temperatures [4]. Water dissolvable or degradable thermoplastic polymers, such as polyglycolic acid (PGA), poly(lactic acid) (PLA),

L. Zhao (✉) · J. Ren · T. Dunne · P. Cheng
CNPC USA Corporation, 2901 Wilcrest Dr, Houston, TX 77042, USA
e-mail: lei.zhao@cnpcusa.com

B. Wisner et al. (eds.), *Composite Materials*, The Minerals, Metals & Materials Series,
https://doi.org/10.1007/978-3-031-50180-7_7

polycaprolactone (PCL), and poly(hydroxybutyrate) (PHB), have been extensively studied and utilized in various applications [5]. For instance, polyvinyl alcohol (PVA) is widely used in water-soluble applications, such as disposable medical gloves and dissolvable substrates in textile manufacturing [6]. Biodegradable PGA is employed in biomedical fields, including implantable medical devices, tissue engineering, and controlled drug delivery [7], and has even been introduced into the oil and gas industry for various downhole equipment [8]. However, these materials have relatively low working temperatures (typically <80 °C) and cannot be utilized as structural materials at high temperatures, despite their readiness to dissolve or degrade in mild, low-temperature conditions [4].

Compared to thermoplastic polymers, thermosets, especially epoxy [9], are more suitable for high-temperature structural applications due to their highly crosslinked chain network structure, resulting in high glass transition temperatures (T_g). Epoxy is favored in the polymer composite industry due to its low-cost, versatile curing chemistries, high mechanical properties, stability, and compatibility with industrial fibers. Although various biodegradable epoxy formulations have been developed and reported, their working temperatures still remain below 100 °C [10]. High-temperature degradable epoxy systems, often investigated for green chemistry, typically necessitate harsh corrosive environments, such as high temperature, high pressure, strong acids/bases, toxic solvents, supercritical fluids, and others [11, 12]. These extreme conditions require capital-intensive and complex equipment or facilities to contain them, leading to high operational costs and energy inputs for maintenance. Apart from the strong epoxy bond, challenges in degrading high-temperature epoxy systems stem primarily from two factors: *strong steric effects* and *low solvent diffusion rates*. The *strong steric effect* arises from functional groups surrounding the degradable bonds within the polymer network [12]. To increase the glass transition temperature and thermal stability, various functional groups (e.g., benzene rings) are introduced to both the epoxy and hardener molecules, limiting the free movement of linkage bonding and neighboring molecular chains. As a result, the degradation of these well-protected ester bonds becomes challenging for water to reach and undergo hydrolysis. The *low solvent diffusion rate* is attributed to the highly condensed and rigid polymer network [13]. Thermosetting materials are known to allow molecular chain rotation and relative movement only above the glass transition temperature, where spaces between the molecular chains provide channels for water molecule diffusion. However, when the temperature falls below the glass transition temperature, the molecular chains "freeze," significantly reducing the entry of water molecules into these channels, leading to an extremely low water uptake and diffusion rate.

Developing a degradable polymer composite system with a high-temperature rating that readily degrades in mild aqueous environments at low temperatures poses a significant challenge, given the conflicting nature of these requirements based on current available techniques. To address this issue and enable large-scale industrial applications, we have chosen industrial-grade epoxy grades and other raw materials, and devised a unique chemistry and microstructure to achieve a high-temperature (>130 °C) degradable epoxy polymer composite that readily degrades in near-neutral aqueous environments at low temperature (≤95 °C). This new technology not only enables the fabrication of workable products for our specific application but also holds immense potential for various other industries, especially in recycling material facilities. By doing so, considerable costs and energy can be saved in current complicated degradation facilities and their operations. Moreover, this approach has the potential to promote the replacement of metals with lightweight polymer composites, contributing to environmental friendliness by reducing carbon emissions during both material manufacturing and improved energy efficiency due to its lighter weight.

Material Fabrication and Characterization

With the aim of enabling large-scale manufacturing and industrial application, all the raw materials, including epoxy, are chosen from commodity products available in industrial grades. To prepare the degradable composite, epoxy with different grades or their mixture is mixed first with acid anhydride hardener. Fine degradation catalyst powder, such as $Ca(OH)_2$, CaO, NaOH, and $Mg(OH)_2$, is then dispersed well in resin mixture with planetary centrifugal mixer under vacuum (FlackTek, DAC 1200 model). Next, half-inch-long E-glass fibers were incorporated into the resin mixture in a planetary mixer, maintaining a 1:1 weight ratio to create bulk molding compound. For molding testing coupons and commercial components, the compound was preheated or precured around 80 °C (so called B-staged) for around 30 min to increase viscosity. The raw materials were then transferred to a compression mold and hot-pressed for 30 min under 6 MPa at around 120 °C. Subsequently, the molded parts underwent post-curing in a vacuum oven at 180 °C for at least 1 h to obtain a glass transition temperature of 205 °C.

A comprehensive investigation of the mechanical, thermal, and dissolving properties of the degradable thermoset composite was conducted to ensure its reliable performance as structural parts. The glass transition temperature (T_g) was measured using a Dynamic Mechanical Analyzer (DMA, TA Instrument Q800 model). The microstructure examination was carried out using a Digital Microscope (Keyence, VHX-7000N). For the material dissolving test in water, molded coupons were soaked in a glass jar at 95 °C. If aged above 100 °C, the test was conducted in an autoclave (Parr Series 4625 High Pressure Vessels). The degradation rate was calculated by dividing the degradation depth by the aging duration.

Results and Discussion

Resin Design and Degradation Mechanism

Hydrolysis below 100 °C is widely regarded as an economical and straightforward degradation method, particularly when contrasted with other conditions involving complex equipment, high-pressure vessels, and potentially hazardous chemicals or solvents. Epoxy resins exhibit a diverse range of crosslinking bonds with various curing agents [9], resulting in ester bonds (with acid anhydrides), ether bonds (with alcohols or phenols), amine bonds (with amine-based hardeners), thioether bonds (with thiol compounds), and urethane bonds (with isocyanates). Among these, only ester bonds and thioether bonds are susceptible to hydrolysis [11]. In this study, ester bonds have been chosen due to the maturity of epoxy/anhydride systems in industrial manufacturing and their application in high-temperature epoxy composites, particularly within the oil and gas industries. This system benefits from extensive research on formulation and mechanical properties, leading to a well-established knowledge base. Moreover, the vast field experience gained from widespread and long-term application enhances the ease of designing with this material and predicting product performance and lifespan.

As mentioned earlier, the hydrolysis rate of high-temperature (HT) epoxy is typically low below its glass transition temperature due to the strong steric effect, low water uptake, and low water diffusion rate [11–13]. To address this issue, it is essential to introduce a degradation catalyst into the polymer matrix system, with better effectiveness achieved by placing it near the ester linkage bond (Fig. 1a, left). This arrangement generates a strong catalytic local environment around the bond, effectively accelerating the hydrolysis process (Fig. 1a). Both acids and bases can serve as excellent hydrolysis catalysts; however, certain considerations must be taken into account when using them in epoxy composites fabrication:

1. Compatibility: The catalyst should not react with the polymer or prepolymer during the mixing and processing stages.
2. Stability: The catalyst should not catalyze hydrolysis prematurely during mixing and processing.
3. Endurance: The catalyst should be able to withstand the processing conditions, including high temperatures, without degradation, vaporization, or loss of effectiveness.
4. Water-free: The catalyst should be water-free to prevent uncontrolled catalyzing effects on epoxy/acid anhydride, which could significantly reduce curing quality.

Considering these requirements, acid or base in fine solid powder form would be the most suitable option. However, solid acid is not a good option for this system due to its rarity, high cost, and typically low melting point (below the curing or post-treatment temperature of HT epoxy) [14]. Our tests have shown that solid acid tends to react easily with unreacted epoxy groups when it melts at elevated temperatures, which not only significantly reduces the material quality but also renders the resin

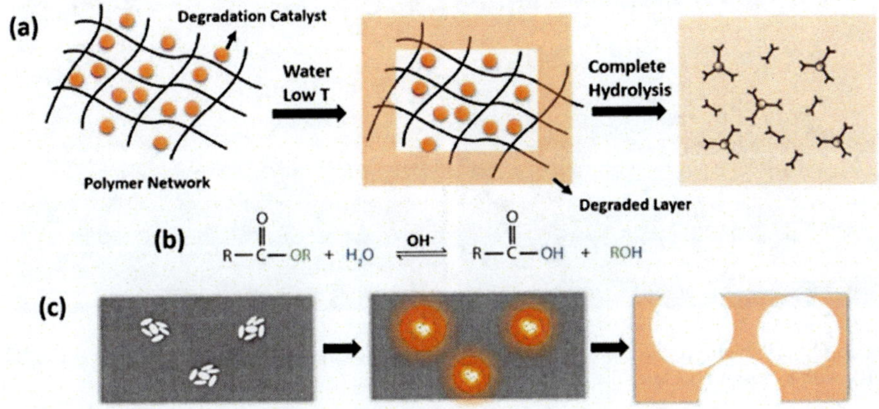

Fig. 1 Schematic illustration of degradable epoxy hydrolysis process (**a**), base-catalyzed hydrolysis reaction (**b**) and explanation of foam microstructure formation mechanism (**c**)

system unable to undergo further hydrolysis (not shown). As a result, solid base fine powders, such as $Ca(OH)_2$, CaO, $NaOH$, and $Mg(OH)_2$, perfectly meet these criteria and are selected as effective and cost-efficient degradation catalysts. The degradation chemistry follows the base-catalyzed hydrolysis reaction process, as shown in Fig. 1b.

To study the detailed degradation process, degradable epoxy resin samples with glass transition temperature of 205 °C are aged at various temperatures, e.g., 95 °C, 110 °C, 130 °C, and 150 °C. In Fig. 2, the surface of the disk shows a clear degradation layer (after aging at 110 °C for 2 days), with the top layer thicker than the bottom due to restricted water access at the bottom side, which is in contact with the container bottom. Dynamic Mechanical Analysis (DMA) tests are conducted on the disc before and after aging, as illustrated in Fig. 3a. The results indicate that aging causes a slight degradation in the mechanical properties of the resin but does not alter its glass transition temperature (T_g). This suggests that water does not diffuse into the undegraded zone. If it did, water molecules could act as plasticizers or swelling agents, facilitating polymer chain movement and reducing the T_g [15]. Two mechanisms are commonly reported to explain polymer degradation dynamics [16]. In the first mechanism, the polymer matrix swells due to the solvent, and degradation occurs throughout the entire polymer network, often observed in rubber degradation. In the second mechanism, degradation gradually occurs from the surface to the interior, primarily due to the limited permittivity of the solvent in the more condensed polymer network. Under this scenario, the material below the degraded surface typically retains its mechanical strength. For this degradable thermoset system, the degradation clearly follows the second mechanism (Fig. 1a, Middle), indicating that the final composite is resistant to moisture and temporary water immersion during field applications. This characteristic makes it particularly suitable for practical applications where exposure to moist environments is a concern [4].

Fig. 2 Optical microscopic cross-sectional image of HT epoxy after 2 days of degradation at 110 °C (left), with a close-up view of the degradation interface area (right)

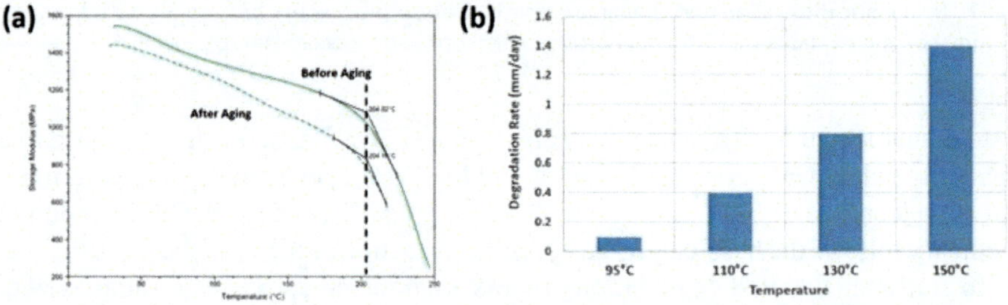

Fig. 3 DMA curves of degradable HT epoxy before and after 2 days of degradation at 110 °C (**a**) and influence of degradation temperature on degradation rate (**b**)

Upon closer examination of the degradation area, a distinct and highly porous foam morphology with interconnected hollow bubbles is evident (Fig. 2, right). Notably, this degradation layer exhibits extreme fragility and can be easily removed by gentle scratching or water turbulence. To elucidate the underlying mechanism, we propose the degradation process illustrated in Fig. 1c. The incompatibility between the organic epoxy resin and the ionic inorganic catalyst solid powders leads to the formation of small segregates, which are uniformly dispersed into the resin matrix (Fig. 1c, left). As water gradually permeates the system, these segregates slowly dissolve, creating a localized, highly alkaline aqueous environment that facilitates the hydrolysis of epoxy ester linkages, ultimately causing degradation. Concurrently, with the dissolution of hydrolysis products, hollow bubbles form and expand during the degradation process (Fig. 1c, middle). As these hollow bubbles merge, they establish continuous pathways to the surrounding aqueous environment, resulting in the complete dissolution and considerable dilution (or removal) of catalyst by the aqueous medium. Consequently, the loss of the degradation catalyst decelerates the degradation process, leaving behind an undegraded skeleton, thus yielding the porous foam structure (Fig. 1c, right). Given these observations, achieving a uniform and high-level dispersion of the catalyst powder is imperative to attain a high level of dissolution. On the other hand, the exceptional fragility of the foam ensures that it merely produces fine organic residue powders when subjected to straining in the degradation tank, thereby avoiding interference with the overall degradation process.

The influence of temperature on the degradation rate, measured by the surface degradation layer thickness over time, is depicted in Fig. 3b. The degradation rate shows a consistent exponential decrease with decreasing temperature, which aligns with previously widely reported ester hydrolysis studies. As a result, degrading high-temperature HT epoxy below 100 °C becomes notably challenging, even with the introduction of a degradation catalyst. At 95 °C, as shown in Fig. 1a, less than 0.1 mm of material is degraded in 1 day, which is considered economically unacceptable for practical applications. Having successfully achieved effective ester hydrolysis in the studied high-temperature system through the introduction of a degradation catalyst, the primary bottleneck restricting the degradation rate lies in the diffusion rate of water molecules into the highly crosslinked and dense epoxy matrix. Therefore, new strategies are required to provide an effective water pathway or "highway" into the material. In response to this challenge, we have evaluated two microstructures: the *bi-continuous microstructure* and the *open-pore microstructures*.

Composite Development Strategy 1: Bi-continuous Microstructure

As previously discussed, degrading high-T_g epoxy at low temperatures poses challenges due to the protective effect of its rigid polymer chains, which shield the degradable ester groups. Additionally, the low water diffusion rate through this highly dense and rigid network further hinders degradation. On the other hand, epoxy with softer polymer chains allows better fluid access to the degradable functional groups, making them more susceptible to hydrolysis. However, these softer epoxy materials typically have very low-T_g values, compromising their mechanical strength at elevated temperatures. To address this issue, we have developed a bi-continuous microstructure by blending compatible high-T_g epoxy with lower T_g epoxy in a carefully chosen ratio, as shown in Fig. 4a. In this design, the high-T_g epoxy forms the backbone of the bulk material, serving as the continuous and dominant phase that provides excellent mechanical strength at elevated temperatures. Meanwhile, the low-T_g epoxy acts as the facilitating phase, contributing to an accelerated degradation rate at lower temperatures. As the degradation progresses and the low-T_g epoxy dissolves, it creates increased open channels or pores within the material. These open channels provide a "highway" for water molecules to diffuse, enabling efficient penetration throughout the material and accelerating the degradation process of the bulk material.

Despite being highly dependent on processing, pure resin typically exhibits weak mechanical properties, with a tensile strength of typically less than 20 MPa, rendering it unsuitable for structural applications. To overcome this limitation and enhance its

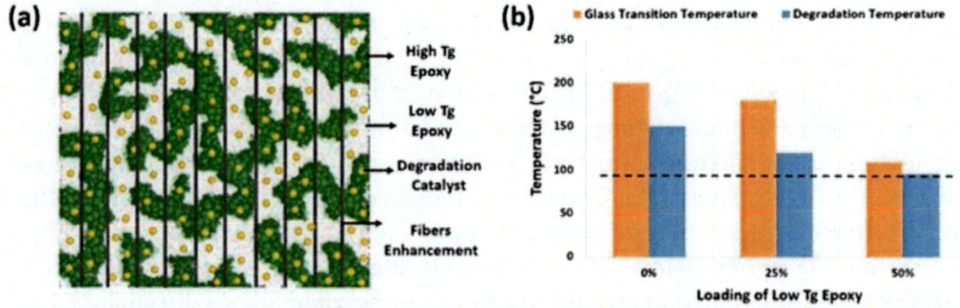

Fig. 4 Schematic illustration of degradable epoxy composite with bi-continuous microstructure (**a**) and influence of low-T_g epoxy loading on both glass transition temperature and degradation temperature of final composites

mechanical performance, compositing with fibers has been demonstrated as a cost-efficient and effective approach [12]. In our previous work, incorporating E-glass fibers led to a significant increase in tensile strength, achieving values of up to 200 MPa [4]. In this study, we introduce chopped glass fibers for practical applications in load-bearing parts (such as hydraulic frac plug used in oil and gas industry), as depicted in Fig. 4a. The addition of these fibers contributes to a substantial improvement in the overall mechanical strength of the composite material, making it suitable for a wide range of practical applications. Moreover, as discussed in the previous section, to achieve the desired degradation behavior, another critical aspect is ensuring the even distribution of the degradation catalyst in this case, throughout the epoxy-fiber mixture.

Various combinations of high-T_g and low-T_g epoxies have been evaluated, but the results have not been satisfactory for this strategy. The properties of one typical combination are shown in Fig. 4b. Without introducing the low-T_g epoxy, the high-T_g epoxy exhibits a T_g around 200 °C. To achieve a dissolving rate of 1 mm/day, a degradation temperature of 150 °C is required. Blending the low-T_g epoxy significantly reduces the required degradation temperature, but it also dramatically lowers the T_g of the mixed epoxy system. The gap between the T_g and the degradation temperature becomes smaller with an increased loading of low-T_g epoxy. As depicted in Fig. 4b, if the required degradation temperature is set at 95 °C, the T_g drops to just above 100 °C. Most low-T_g epoxy has aliphatic and flexible polymer chains, which act as plasticizers to the high-T_g epoxy with rigid aromatic chains [5, 6]. Consequently, it is challenging to maintain a high T_g of the overall system with an increased loading of the low-T_g component to form effective connected open channels or pores. This situation poses a challenge in finding an optimal balance between maintaining a high T_g and achieving the desired low degradation temperature.

Composite Development Strategy 2: Open-Pore Microstructure

In this design (Fig. 5a), a rigid and highly condensed high-T_g epoxy network is deliberately maintained to preserve its high-temperature rating, while open pores or channels are intentionally introduced to facilitate high water diffusion. Similar to the previous strategy, short glass fibers are also incorporated to enhance mechanical strength, and a degradation catalyst is introduced into the polymer matrix to enable its degradation property. To achieve these open pores or channels, a unique formulation and processing method has to be followed: one particular industrial-grade high-T_g epoxy monomer, with a boiling point below 180 °C, is selected and blended with an acid anhydride hardener in a specific ratio that is insufficient to react with all the epoxy groups. The system is first cured at a low temperature (typically between 100 and 120 °C) to form high-quality parts, which are then aged at 180 °C inside an industrial vacuum oven. This aging process, involving high temperature plus vacuum, efficiently evaporates the unreacted high-T_g epoxy monomer, resulting in uniform and effective open channels within the polymer matrix. In one representative formulation, the degradable composites exhibit a high glass transition temperature around 130–140 °C (Fig. 5b), and they can readily dissolve in water at 95 °C with a degradation rate as high as 2.5 mm/day (Fig. 6). This unique combination of materials and processing techniques allows the material to maintain its high T_g for high-temperature applications while enabling a high degradation rate at low temperatures.

In this innovative technique, the initial ratio of hardener to epoxy monomer plays a pivotal role in determining the properties of the degradable composite. The results obtained from various ratios are effectively summarized in Table 1. When the hardener ratio is too low, it prevents the formation of a well-consolidated epoxy network, leading to a reduced glass transition temperature (T_g) [17]. Within this particular range, T_g increases with the hardener ratio as crosslinking is enhanced [18]. Conversely, if the hardener ratio is too high (e.g., 1:1.8 in Table 1), all the epoxy groups become fully crosslinked within the network, leaving no free epoxy monomers to generate open pores or channels during the subsequent post-aging procedure. As a result, the degradable composite loses its ability to degrade at low temperatures,

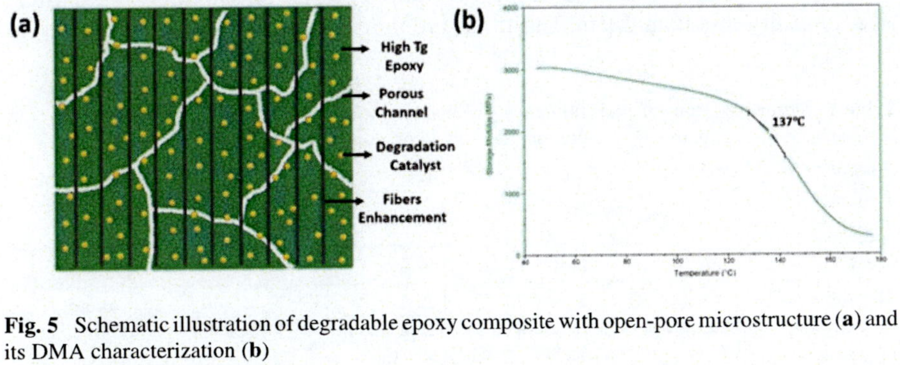

Fig. 5 Schematic illustration of degradable epoxy composite with open-pore microstructure (**a**) and its DMA characterization (**b**)

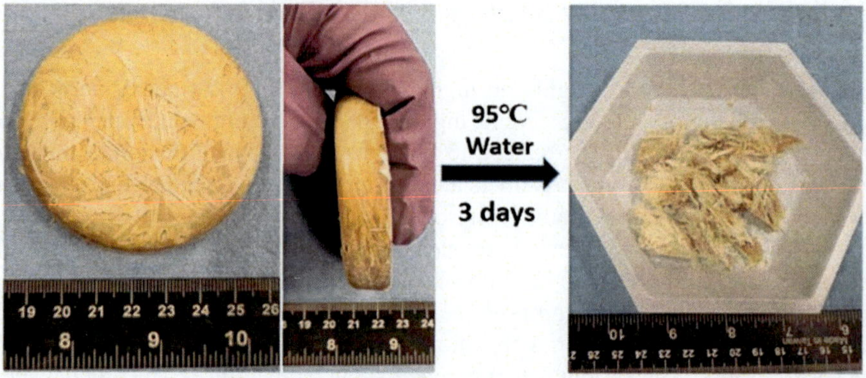

Fig. 6 Photographic images of the degradable epoxy composite before and after aging in 95 °C water for 3 days

behaving similarly to the HT system observed in Fig. 2. To achieve the desired properties, it is critical for the epoxy/hardener ratio to fall within a specific narrow range, enabling the establishment of an optimal balance between high-temperature performance and low-temperature degradation rate. The representative example depicted in Figs. 5b and 6 demonstrates that this favorable range lies between 1:1 and 1:1.4 (Table 1).

In addition to the epoxy/hardener ratio, another crucial parameter that significantly influences the quality and degradation rate of the components is the loading of the degradation catalyst as observed in the same system discussed in Fig. 6. This material's performance under different loadings is elucidated in Fig. 7a. Consistent with fundamental mechanism studies, the degradation rate of the overall composite increases with higher catalyst loading. To ensure effective degradation below 100 °C, a minimum loading of 40 wt% (relative to the resin) is necessary. Moreover, employing a high catalyst loading serves a dual purpose as it acts as a filler to enhance the resin's viscosity. This, in turn, contributes to the successful molding of high-quality parts through typical industrial compressive molding processes. To achieve this, it is essential to use the degradation catalyst in an extremely fine particle size, typically less than 2 μm, and to ensure thorough blending with the epoxy to

Table 1 Summarization of the influence of epoxy monomer/hardener weight ratio on the performance of degradable epoxy composite with an open-pore microstructure

Epoxy/ hardener wt. ratio	1:0.5	1:0.7	1:1	1:1.4	1:1.8
Result	Fail	Fail	Work	Work	Fail
Reason	T_g < 110 °C Degrade at 95 °C	T_g < 130 °C Degrade at 95 °C	T_g > 130 °C Degrade at 95 °C	T_g > 140 °C Degrade at 95 °C	Can not degrade at 95 °C

(a)

Catalyst Loading	Result	Reason
17%	Fail	Non-degrade at 95°C
28%	Fail	Non-degrade at 95°C
44%	Work	degrade at 95°C
54%	Work	degrade at 95°C
64%	Fail	Too Thick to Process

(b)

Fig. 7 Summarization of the influence of degradation catalyst loading on the performance of degradable epoxy composite with an open-pore microstructure (**a**) and photographic image showcasing a high-quality commercial component made from this material (**b**)

mitigate particle agglomeration. However, it is essential to avoid excessive catalyst loading as it could lead to the resin mixture becoming too thick to efficiently blend with chopped glass fibers in the subsequent step. Our research, as presented in Fig. 7a, indicates that a catalyst loading of 64 wt.% represents the upper limit for industrial processing to obtain bulk molding compound. Beyond this threshold, the practicality and effectiveness of the molding process may be compromised [19]. Furthermore, during the molding process, pre-crosslinking (B-stage) of the compound is also necessary. This step further enhances the molding process's efficiency and overall part quality [20]. By combining and synergizing these techniques effectively, a high-quality part with the formulation depicted in Fig. 6 has been successfully achieved and employed in our high-temperature dissolvable plug applications (Fig. 7b).

Conclusions

This study presents a pioneering solution in the form of a novel high-T_g (>130 °C) epoxy composite that exhibits water degradability below 100 °C. It addresses the challenges associated with existing degradable materials, which either have low working temperature ratings (such as thermoplastics and biodegradable epoxy) or require extremely harsh and corrosive conditions to degrade despite their high-temperature ratings. Through the successful achievement of effective ester hydrolysis in the high-temperature system by introducing a degradation catalyst and enhancing water diffusion with open micro-channels, we have overcome the limitations restricting the degradation rate, enabling efficient degradation even at low temperatures.

Moreover, this study thoroughly investigates and illustrates the degradation mechanism of the technique, systematically exploring various parameters, including catalyst selection and loading, epoxy/hardener ratios, etc., to establish the microstructure–fabrication–property relationships of this new system. This knowledge provides a platform for the material science community to develop more advanced material systems with enhanced properties. The innovative approach not only opens up possibilities for diverse industrial applications but also offers significant potential for the recycling material industry, resulting in substantial cost and energy savings, along with enhanced environmental friendliness. As a result, it aligns perfectly with the current industrial trend of replacing traditional metals with lightweight composites, promoting energy efficiency and contributing to carbon reduction efforts.

Acknowledgements The authors thank the management of CNPC USA for granting permission to publish this work.

References

1. Zhao L, Dunne TR, Ren J, Cheng P (2023) Dissolvable magnesium alloys in oil and gas industry. In: Tański TA (ed) Magnesium alloys—processing, potential and applications. IntechOpen, London, UK
2. Xu ZY, Zhang ZH (2019) The art of disintegration–ten years in review of disintegrable metals and downhole tools. Paper presented at Offshore technology conference, Houston, USA, May 2019
3. Okura M, Takahashi S, Kobayashi T, Saijo H, Takahashi T (2015) Improvement of impact strength of polyglycolic acid for self-degradable tools for low-temperature wells. Paper presented at SPE Middle East unconventional resources conference and exhibition, Muscat, Oman, January 2015
4. Zhao L, Ren J, Yuan M, Liu Y, Liu HL, Cheng P (2023) Novel hybrid degradable plugs to enable acid fracturing at high temperature high pressure conditions. In: Paper presented at SPE annual technical conference and exhibition, San Antonio, USA, October 2023
5. Nair LS, Laurencin CT (2007) Biodegradable polymers as biomaterials. Prog Polym Sci 32:762–798
6. Huang D, Hu ZD, Ding Y, Zhen ZC, Lu B, Ji JH, Wang GX (2019) Seawater degradable PVA/PCL blends with water-soluble polyvinyl alcohol as degradation accelerator. Polym Degrad Stab 163:195–205
7. Samantaray PK, Little A, Haddleton DM, McNally T, Tan B, Sun ZY, Huang WJ, Ji Y, Wan CY (2020) Poly(glycolic acid) (PGA): a versatile building block expanding high performance and sustainable bioplastic applications. Green Chem 22:4055–4081
8. Takahashi S, Shitsukawa A, Okura M (2018) Degradation study on materials for dissolvable frac plugs. In: Paper presented at the SPE/AAPG/SEG unconventional resources technology conference, Houston, US, July 2018
9. Ellis B (ed) Chemistry and technology of epoxy resins. Springer Science+Business Media, Berlin
10. Shen M, Almallahi R, Rizvi Z, Martinez EG, Yang GZ, Robertson ML (2019) Accelerated hydrolytic degradation of ester-containing biobased epoxy resins. Polym Chem 10:3217–3229
11. Liu T, Guo XL, Liu WC, Hao C, Wang LW, Hiscox WC, Liu CY, Jin C, Xin J, Zhang JW (2017) Selective cleavage of ester linkages of anhydride-cured epoxy using a benign method and reuse of the decomposed polymer in new epoxy preparation. Green Chemistry 19(18):4364–4372
12. Oliveux G, Dandy LO, Leeke GA (2015) Current status of recycling of fibre reinforced polymers: review of technologies, reuse and resulting properties. Prog Mater Sci 72:61–99
13. Wong TC, Broutman LJ (2004) Water in epoxy resins part II. Diffusion mechanism. Polym Eng Sci 25(9):529–534
14. Clark JH (2002) Solid acids for green chemistry. Acc Chem Res 35:797–797

15. Yuan YS, Goodson J (2008) HTHP in-situ mechanical test rig and test method for high-temperature polymers and composites. Paper presented at SPE Europec/EAGE annual conference and exhibition, Rome, Italy, June 2008
16. Gopferich A (1996) Mechanisms of polymer degradation and erosion. Biomaterials 17(2):103–114
17. Wu LX, Hoa SV, Tan M, That T (2006) Effects of composition of hardener on the curing and aging for an epoxy resin system. J Polym Sci 99:580–588
18. Bignotti F, Pandini S, Baldi F, Santis RD (2011) Effect of the resin/hardener ratio on curing, structure and glass transition temperature of nanofilled epoxies. Polym Compos 32:1034–1048
19. Davis BA (2003) Compression molding. Hanser, Liberty Township, OH, p 49
20. Studer J, Dransfeld C, Masania K (2016) An analytical model for B-stage joining and co-curing of carbon fibre epoxy composites. Compos A Appl Sci Manuf 87:282–289

Part II
Eco-Friendly and Sustainable Composite Materials: Waste Stream Benefits

Material Developments for 3D/4D Additive Manufacturing (AM) Technologies

Simona E. Hunyadi Murph and Henry T. Sessions Jr.

Abstract Additive Manufacturing (AM), or 3D printing, is a unique technology in which structurally complex objects can be easily manufactured. While AM allows for the creation of intricate 3D objects, these objects are inactive and motionless. With recent incorporation of a "pre-programmed functionality" into the 3D printed objects, a new concept has emerged, 4D printing. In this context, the pre-programmed functionality refers to the materials that have properties related to their electrical, magnetic, optical properties, etc., that can be manipulated in predictable ways by application of external stimuli. Therefore, the 4D printing technology enables a static 3D printed object to change its shape, functionality, or property over time upon exposure to specific stimuli such as heat, stress, light, pH, moisture, etc. We describe a variety of functional composite (4D) materials developed for incorporation in hydrogen storage-based applications.

Keywords Additive manufacturing · Hydrogen storage materials · Functional 3D/4D materials

Introduction

The additive manufacturing (AM) technology, or 3D printing, is one of the fastest growing sectors in the market with an ascendent growth for over two decades [1]. The estimated global market for AM was approximately $21 billion in 2020, and it is expected to reach a value of $63.46 billion by 2026 [2]. AM is a unique technology in which structurally complex objects can be easily manufactured. Using 3D modeling and computer-aided design (CAD), customized parts and components are produced predictably through a layer-by-layer deposition approach.

S. E. Hunyadi Murph (✉) · H. T. Sessions Jr.
Savannah River National Laboratory, Aiken, SC 29808, USA
e-mail: Simona.Murph@srnl.doe.gov

S. E. Hunyadi Murph
University of Georgia, Athens, GA 30602, USA

B. Wisner et al. (eds.), *Composite Materials*, The Minerals, Metals & Materials Series,
https://doi.org/10.1007/978-3-031-50180-7_9

AM methods have several advantages over traditional manufacturing techniques. For example, the AM offers a "design freedom" that allows creation of structurally complex objects that are unbuildable in a traditional machine shop. The AM process reduces waste generation and eliminates the need for additional etching and cleaning steps, which minimizes use of harmful chemicals. Furthermore, with AM it is possible to create parts and components without the need for assembly. However, while AM allows for the creation of intricate 3D objects, to date, the large majority of AM printed objects serve as structural components only for devices. These objects are inactive, "stationary and motionless" [3, 4].

In recent years, exploration of functional materials for AM has emerged, bringing a new level of excitement into the scientific community. It is possible for 3D printed parts to not only serve as structural components, but also serve as functional components [1]. The idea of incorporating "pre-programmed functionality" into 3D printed objects coined a new phrase, "4D printing". The pre-programmed functionality expands the structural material properties through the addition of a distinct "functionality" associated with properties such as electrical, magnetic, thermal, or optical, among others. Specifically, the 4D printing technology enables a static 3D printed object to change its shape, functionality, or property over time upon exposure to specific stimuli such as heat, stress, light, pH, and moisture. As a result, 4D printing has become an exciting branch of additive manufacturing and attracts significant interest from various stakeholders [1].

While the 4D printing is a new, relatively unexplored arena, the field is rich with opportunities for generation of exquisite parts with distinctive property gradients and features. To date, two types of active materials have been under investigation for use in 4D printing: shape memory polymers and hydrogels [5]. Recently, 4D printing of metals using nickel–titanium shape memory alloy via multistage transformation process was reported [6]. In order to take advantage of additive manufacturing's full potential in the materials domain, key breakthroughs in the research, development, and integration of functional materials are needed. The focus should not only be on the manufacturing process, but also on the in situ and post-processing characterization and evaluation of AM processes and functional materials.

The field is rich with opportunities for innovations at various scales and processes. Developing high-quality feedstock materials with functional capabilities at scale for targeted applications is at the forefront. To accelerate and enable the development of functional materials for additive manufacturing processes, a focus must be placed on the production of multicomponent materials with functional properties, i.e. piezoelectric, thermoelectric, biologic, magnetic, semiconductor, optical, and superconductivity among others [7–11]. AM of functional materials is not trivial. Functional components require a high degree of control over material properties (crystallinity, grain size, defects, grain boundaries, size, shape, orientation, etc.) at various scales (nano/micro/meso/macro-scale). Accurate and strategic deposition at desired location, post-processing effects, and the ability to be appropriately manipulated by external stimuli are a few examples of parameters needed to be taken under consideration during AM process.

Hydrogen is of increasing interest as a clean fuel for fuel cells, pumping and compression systems, absorption coolers, etc. [12]. One particular area of interest that can greatly benefit from the implementation of the 3D/4D printed objects is in hydrogen fuel cells for electric vehicles [13]. Hydrogen fuel cells utilize proton exchange membrane (PEM) technology to generate electricity from stored hydrogen and atmospheric oxygen. Batteries are still the main source of stored energy for electric vehicles and battery technology continues to improve. Weight, energy capacity, and recharging times for batteries (even for the popular lithium–ion batteries) still limit broad adoption. Hydrogen fuel cells (HFC) have been investigated for use in electric-powered vehicles for years. Moreover, electric motors are more efficient than gas-powered internal combustion engines. Gasoline-fueled internal combustion engines are typically 35% efficient while electric vehicles are over 65% efficient. Electric vehicles are quiet and produce no environmentally harmful emissions.

A major drawback to the wider utilization of hydrogen and fuel cells for vehicle power is the lack of safe and efficient hydrogen storage mediums. To take advantage of the AM's full potential in the materials domain, key breakthroughs in the research, development, and integration of functional materials are needed. Therefore, we report the design and production of hydride-based 3D/4D printable functional materials, namely, Mg-based and Pd-based materials, that can be used as hydrogen storage materials for use as structural component incorporation in the hydrogen storage-based applications.

Experimental

Magnesium–Aluminum–Silicon powder feed stock was used to produce 3D printed components using a Concept Laser X-1000R printer, with parameters developed for a selective laser melting process. The feed stock material, Ti6Al4V, with a diameter range of 45–106 μm was purchased from Advanced Powders and Coatings. The laser sintering 3D printing uses a high-power laser to fuse metallic powdered material by scanning cross sections on the surface of a powder bed. This is a preferred technique for the Ti6Al4V materials. The printer specifications included: laser system: 1000W Yb-fiber, focus diameter: 100–500 μm, build plate heating: up to 200 °C, inert atmosphere: argon. Printed Ti6Al4V materials scaffolds were manufactured via powder bed EBM with an Arcam EBM A2X printer from Arcam AB (Mölndal, Sweden) of General Electric. A Hitachi 4800 scanning electron microscope, with composition analysis detection, was used to obtain electron microscopy data. The porous iron samples were purchased from Hoganas, North America and the Fe_2O_3 nanoparticles were purchased from Sigma-Aldrich. Pd nanoparticles were produced by reduction approaches, as previously reported by SRNL, using Fe_2O_3 NPs as seed supports. Isotherms were collected at room temperature (294 K) with a custom in-house built Sieverts apparatus described earlier [8].

Results and Discussion

Hydrogen fuel can be stored in either: (1) gas phase under high Pressure, (2) in the liquid phase at extremely low temperatures, or (3) solid state under low-pressure hydride materials. High-pressure hydrogen storage vessels are bulky, heavy, and pose a safety concern, especially for use in passenger vehicles. Low-temperature liquid phase storage is even less feasible for use in electric-powered vehicles as it requires expensive processing and storage facilities. The best alternative for the safe storage of hydrogen, for subsequent use with PEM fuel cells to power electric vehicles, relies on the use of solid-state hydride materials that are lightweight and able to store hydrogen at low pressure, and can release the stored hydrogen in an easily controllable manner.

Hydrogen storage in magnesium (Mg) has been the subject of an intense research effort in the last decades due to its favorable properties [14–16]: (i) high theoretical gravimetric hydrogen density of 7.6 wt%, (ii) reversibility, (iii) low cost, and (iv) natural abundance, e.g. eighth most abundant element in the earth's crust (2.3%) and the third most abundant element dissolved in sea water. However, the rate of hydrogenation is slow at moderate temperatures (200–300 °C), and it requires rather high temperatures (>400 °C) to achieve complete conversion of Mg to MgH_2 at a reasonable rate [17]. In recent years, researchers have sought to better engineer or allowing magnesium with other metals and catalysts, at either nano- and macro-scales to improve their thermodynamic and kinetics response. For example, several studies reported significant uptake of hydrogen by Mg *at room temperature* [9, 11, 12] by alloying with a number of additive components such as Pd [18], TiF_3 [19], and Ti–Zr [20]. Other studies reported that alloying Mg with transition metals Sc, Zr, Ni, Co, Mn, Fe, or V provide substantial improvements at moderate temperatures of their the activation energy, heat transfer, and the kinetics [21]. It was also found that that the addition of ca. 15% of Al decreases the binding energy to the hydrogen to the target value of 0.25 eV which corresponds to release of 1 bar hydrogen gas at 100 °C [22].

Functional Magnesium-Based Hydride Materials

Designing 3D printed structural components for hydrogen fuel cell-powered electric vehicles requires the deposition of an active metal hydride encased in an inert metal that serves as a scaffolding, i.e. aluminum having a shape of a structural component of a vehicle or a portion thereof. Therefore, Mg-based hydride material with aluminum scaffolding was designed and printed in a defined architecture using a Concept Laser X-1000R parameter development with the selective laser melting process. The 3D/4D printed materials were subsequently characterized by a series of analytical tools to evaluate their physico-chemical properties (Fig. 1).

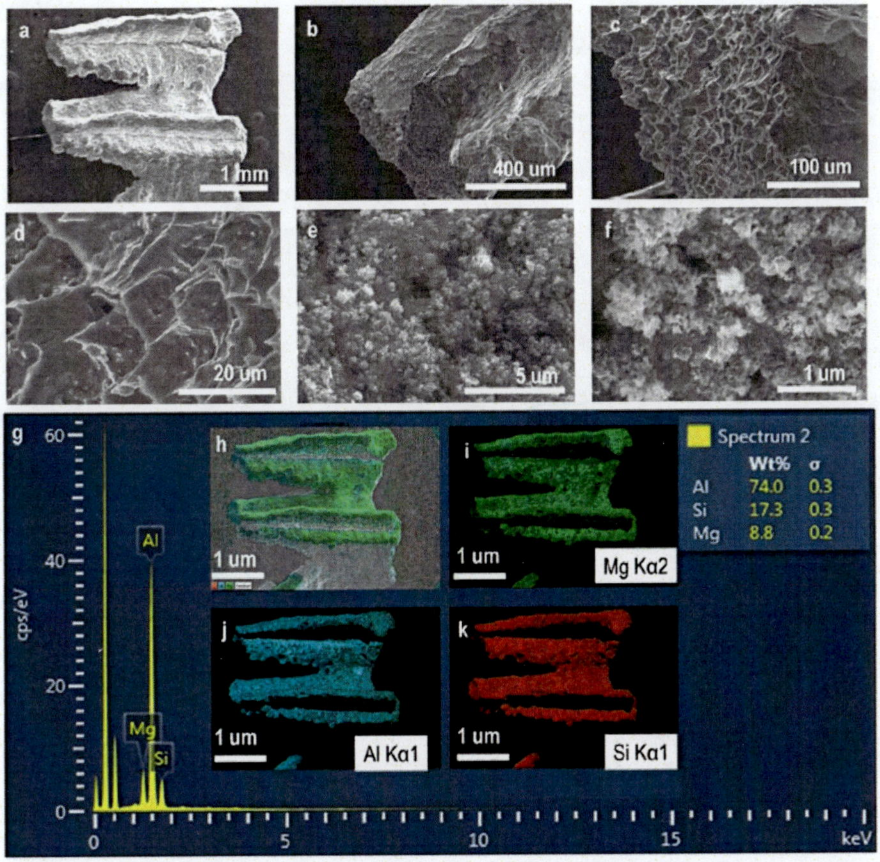

Fig. 1 **a–f** Different resolution scanning electron microscopy images (from 1 mm to 1 μm scale) collected on the Mg–Al–Si 3D printed component showing the morphology and crystallite and pore-like structures; **g** EDS data showing compositional identity of the material (Wt%: Al 74%, Si 17.3%, 8.8%); **h–k** EDS mapping of individual material

The morphological and structural studies performed on the resulting 3D printed Mg-based materials reveal that this particular composition can be manufactured in a variety of conformations and porosities. Specifically, large structures of complex shapes can be produced. Scanning electron microscopy analysis conducted at different resolutions (Fig. 1a–f) shows production of uniform and quasi-smooth side surfaces. The 3D printed material, however, displays tubular (approximately 10 μm in diameter), porous-like structures and protuberances at the end of the structure. To elucidate the compositional configuration of the material printed, a series of X-ray energy-dispersive microanalysis (EDS) was performed. Data shows that the materials retain their properties upon 3D additive manufacturing. EDS mapping (Fig. 1g–k) shows uniform distribution of Mg on the entire 3D printed component. This is not surprising, EDS data (Fig. 1) reveal that approximatively 9% Mg was present with

manufacturing or deposition of nanomaterials post production of the manufacturing process. Palladium (Pd) nanoparticles have favorable functional properties that make them valuable for hydrogen storage application, energy-related application, fusion, or separations technologies [18, 28–31]. Pd absorbs large volumetric quantities of hydrogen at room temperature and atmospheric pressure forming palladium hydride, exhibiting rapid sorption kinetics, favorable thermodynamic properties, and long-term cycling stability [32]. Hydrogen stored in palladium material can be absorbed and released and used to fuel the electric vehicles through careful temperature and pressure control [33]. Palladium nanoparticles are produced through the reduction of molecular precursors [34] on porous iron and iron oxide nanospheres.

The reaction that decorates iron-based components with palladium nanostructures is based on a metal ion reduction procedure reported by us earlier [35]. By employing these experimental conditions, one could preserve the structural integrity of the support, which is critical in this study. A complete characterization approach, i.e. SEM, EDS, EDS mapping, was conducted to elucidate the deposition patterns and materials' properties.

The porous iron has a sponge-like porous structure (Fig. 3) with a BET surface area of 0.95 m^2/g, pore volume of 0.0068 mL/g, and average pore diameter of 364 Å. It is >92% amorphous and α-iron with small amounts of Fe and Al oxides [36]. Fe_2O_3 nanospheres, with a diameter of 70 ± 37 nm, were also decorated with palladium nanoparticle. In both cases, Pd quasi-spherical nanostructures with a diameter of 27 ± 10 nm were fairly uniform decorated on the iron-based components. Upon decoration with Pd, the magnetic properties of the iron-based materials remained intact.

The behavior and quality of a metal hydride material is commonly assessed by pressure–composition–temperature (PCT) isotherms. Hydrogen absorption/desorption by responsive materials is a reversible process. The generation of useful isotherm data requires a leak tight temperature-controlled apparatus and a leak tight vessel for the material being tested. During hydrogen absorption, heat must be removed from the vessel as the absorption reaction is exothermic. During the desorption reaction, however, heat must be removed from the system as the desorption reaction is endothermic.

The absorption process starts by introduction of a known amount of hydrogen, an aliquot, into the vessel containing the material under test. The vessel is maintained at a controlled temperature and the vessel pressure is constantly monitored. This pressure data and known quantity of material under test allow for the calculation of the atomic ratio of hydrogen and absorbing material (H/M). The pressure versus H/M is plotted to generate the isotherm. The initial introduction of small aliquots of hydrogen generates a metal hydride, PdH_x, known as alpha phase [7, 8, 30]. Subsequently, as more hydrogen is interstitially incorporated, the material transitions from alpha to beta phase. This is indicated by the plateau of the isotherm and the plateau is also an indication of the materials' storage capacity. The right portion of the curve indicates a pressure increase in the vessel and thus the limit of the material's hydrogen storage capacity. Figure 3c shows an example of the PCT absorption isotherm for palladium nanoparticles with <25 nm in diameter at

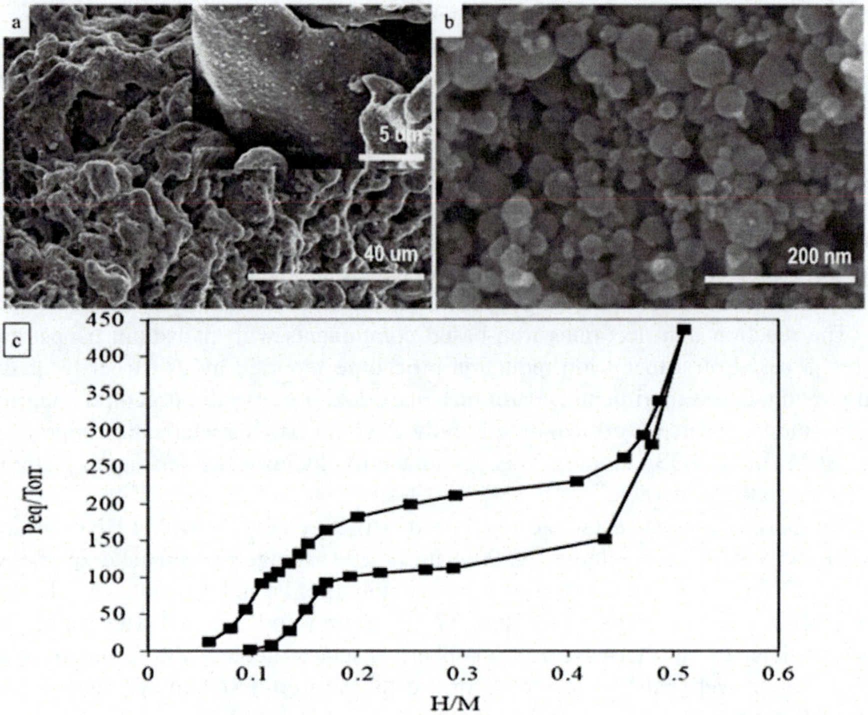

Fig. 3 **a** SEM images of porous iron-Pd composite, **b** SEM image of Pd–Fe_2O_3 composite, **c** representative hydrogen absorption and desorption isotherm collected at 90 °C on Pd nanoparticles of <25 nm in diameter

90 °C. The calculated capacity for palladium is 0.5 H/M for hydrogen. The sorption capacity can be fine-tuned by controlling the particle size. A smaller size promotes higher kinetics of gas uptake and release. In the same time, the diffusion process can be tailored by the material's morphology and dimension. A porous structure with smaller dimensions boosts the hydrogen exchange by shortening diffusion distances. Our studies showed that coupling palladium with other materials increases the materials' sorption capacity [7, 8, 27, 30]. For example, the plateau pressure at room temperature for a hybrid Fe_2O_3–Pd nanoparticles ($d = 27$ nm) was ≈10 Torr, with a calculated capacity of 0.55 H/M for hydrogen. This demonstrates that the functional material retains its hydride properties when manufactured on other media. The bulk hydride materials often decrepitate or break down into smaller particles under repetitive reaction cycles, usually hundreds of cycles. These results support the notion that incorporation of Pd nanoscale materials with favorable hydrogen gas storage capabilities is advantageous when manufacturing 3D functional hydrogen storage materials. Additional post-processing studies should be conducted to capture materials' behavior during the printing or post-processing procedure to ensure long-term performance.

Conclusions

We demonstrated production of materials that can be used or incorporated as hydrogen storage components for use as structural components in conjunction with hydrogen fuel cells in electric-powered vehicles. The hydrogen storage materials can be in the form of a 3D printed Mg-based metal alloys or Pd-based nanoscale-enabling materials. These materials retain their behavior and properties upon manufacture. If incorporated in devices, we foresee that the structural components that carry these hydrogen storage materials could serve as viable exchangeable components of a vehicle.

Acknowledgements This work was supported by the Laboratory Directed Research and Development (LDRD) program within the Savannah River National Laboratory (SRNL). This work was produced by Battelle Savannah River Alliance, LLC under contract no. 89303321CEM000080 with the U.S. Department of Energy. Publisher acknowledges the U.S. Government license to provide public access under the DOE Public Access Plan (http://energy.gov/downloads/doe-public-access-plan).

References

1. The Minerals, Metals & Materials Society (TMS) (2022) Accelerating research and technological development in the additive manufacturing of energy-related functional materials (2022). www.tms.org/AMStudy
2. Mordor intelligence: 3D printing market size, trends-industry growth 2022–27
3. Mitchell A et al (2018) Additive manufacturing—a review of 4D printing and future applications. Addit Manuf 24:606
4. Yang C et al (2019) 4D printing reconfigurable, deployable and mechanically tunable metamaterials. Mater Horiz 6:1244–1250
5. Wei H (2017) Direct-write fabrication of 4D active shape-changing structures based on a shape memory polymer and its nanocomposite. ACS Appl Mater Interfaces 9:876
6. Caputo MP et al (2018) 4D printing of net shape parts made from Ni–Mn–Ga magnetic shape-memory alloys. Add Manuf 21:579
7. Hunyadi Murph SE et al (2021) Efficient thermal processes using alternating electromagnetic field for methodical and selective release of hydrogen isotopes. Energy Fuels 35:3438–3448. https://doi.org/10.1021/acs.energyfuels.0c03704
8. Hunyadi Murph SE (2020) Shape-selective mesoscale nanoarchitectures: preparation and photocatalytic performance. Catalysts 10:532. https://doi.org/10.3390/catal10050532
9. Hunyadi Murph SE et al (2017) Nanoparticle treated stainless steel filters for metal vapor sequestration. JOM 69:162–172. https://doi.org/10.1007/s11837-016-2206-5
10. Larsen G et al (2016) Multifunctional Fe_2O_3–Au nanoparticles with different shapes: enhanced catalysis, photothermal effects, and magnetic recyclability. J Phys Chem C 120:15162–15172. https://doi.org/10.1021/acs.jpcc.6b03733

11. Hunyadi Murph SE (2022) Shape-selective palladium and palladium-composite nanomaterials. In: Metal-matrix composites. Springer Nature, pp 227–245. https://doi.org/10.1007/978-3-030-92567-3_15
12. Hunyadi Murph SE, Sessions H et al (2023) Automatic gas sorption apparatus and method. US11307129B. https://patents.justia.com/patent/11307129
13. Hunyadi Murph SE et al (2021) Structural vehicle components for hydrogen storage. US Patent App. 16/822,634. https://patents.justia.com/patent/20210291267
14. Edalati K et al (2018) Design and synthesis of a magnesium alloy for room temperature hydrogen storage. Acta Mater 149:88–96
15. Bolarin JA et al (2021) Room temperature hydrogen absorption of Mg/MgH_2 catalyzed by $BaTiO_3$. J Phys Chem C 125(36):19631–19641
16. Baran A, Polański M (2020) Magnesium-based materials for hydrogen storage—a scope review. Materials 13(18):3993. https://doi.org/10.3390/ma13183993. PMID 32916910; PMCID PMC7559164
17. Lu J et al (2010) Hydrogenation of nanocrystalline Mg at room temperature in the presence of TiH_2. J Am Chem Soc 132:6619
18. Gautam YK et al (2015) A room temperature hydrogen sensor based on Pd–Mg alloy and multilayers prepared by magnetron sputtering. J Int Hydrogen Energy 40:15549
19. Xie L et al (2007) Superior hydrogen storage kinetics of MgH_2 nanoparticles doped with TiF_3. Acta Mater 55:4585
20. Chen C et al (2018) Improved kinetics of nanoparticle-decorated Mg–Ti–Zr nanocomposite for hydrogen storage at moderate temperatures. Mater Chem Phys 206:21
21. Schneemann A et al (2018) Nanostructured metal hydrides for hydrogen storage. Chem Rev 118:10775
22. Andreasen A (2018) Hydrogenation properties of Mg–Al alloys. J Int J Hydrogen Energy 33:7489
23. Hunyadi Murph SE et al (2017) Anisotropic and shape-selective nanomaterials: structure-property relationships. Springer Nature. https://doi.org/10.1007/978-3-319-59662-4
24. Tao CG et al (2007) Surface morphology and step fluctuations on silver nanowires. Surf Sci 601:4939–4943. https://doi.org/10.1016/j.susc.2007.08.023
25. Hunyadi Murph SE et al (2013) Patchy silica-coated silver nanowires as SERS substrates. J Nanoparticle Res 15(6):1607. https://doi.org/10.1007/s11051-013-1607-4
26. Hunyadi Murph SE et al (2012) Metallic and hybrid nanostructures: fundamentals and applications. Appl Nanomater. In: Govil JN (ed) Nanomaterials and nanostructures, vol 4. Studium Press LLC, USA.
27. Hunyadi Murph SE, Larsen G, Lascola R (2016) Multifunctional hybrid Fe_2O_3–Au nanoparticles for efficient plasmonic heating. J Vis Exp (JOVE) 108:e53598. https://doi.org/10.3791/53598
28. Hunyadi Murph SE et al (2011) Synthesis, functionalization, characterization and application of controlled shape nanoparticles in energy production. In: Fluorine-related nanoscience with energy applications. ACS symposium series, vol 1064, chap 8, pp 127–163. https://doi.org/10.1021/bk-2011-1064.ch008
29. Hunyadi Murph SE, Schyck S, Bass J (2022) Shape-selective palladium and palladium-composite nanomaterials. In: Metal-matrix composites. Springer Nature, pp 227–245
30. Srivatsan TS, Harrigan WC, Hunyadi Murph SE (2021) Metal-matrix composites. Advances in analysis, measurement, and observations, p 281. https://doi.org/10.1007/978-3-030-65249-4
31. Hunyadi Murph SE et al (2022) Engineered nano-antenna susceptor as efficient platforms for efficient uptake and release of analytes. In: Srivatsan TS, Rohatgi PK, Hunyadi Murph S (eds) Metal-matrix composites. TMS. Springer, Cham. https://doi.org/10.1007/978-3-030-92567-3_22
32. Hunyadi Murph SE, Sessions H et al (2021) Energy and Fuels 35:3438–3448
33. Hunyadi Murph SE, Murph MA (2022) Nuclear fusion: the promise of endless energy. Phys Sci Rev. https://doi.org/10.1515/psr-2021-0069

34. Hunyadi Murph SE et al (2021) Systems and methods for manufacturing nano-scale materials. US20210380405A1
35. Hunyadi Murph SE et al (2020) Controlled release of hydrogen isotopes from hydride-magnetic nanomaterials. ACS Appl Mater Interfaces 12(8):9478–9488
36. Li D et al (2019) Porous iron material for TcO_4^- and ReO_4^- sequestration from groundwater under ambient oxic conditions. J Hazard Mater 374:177–185

Influence of the Use of Anti-bubble Additives on the Permeability and Porosity of Anticorrosive Coatings

D. C. R. Velasco, D. L. R. Oliveira, F. P. D. Lopes, D. Souza, and C. M. F. Vieira

Abstract The demand for anticorrosive protection materials is well established, and the utilization of industrial pipe coatings presents a viable solution. Within this realm of protection, the utilization of particulate composites is common, as it facilitates the establishment of a protective barrier between the pipe and the corrosive environment. The objective of this study is to evaluate the feasibility of integrating additives to reduce the permeability and porosity of pipe coatings. A water vapor permeability test was conducted in accordance with ASTM D1653 (Method B-Condition C), while porosity was assessed following the guidelines of NACE SP0394. The evaluated coating uses organic and/or inorganic particulate waste. Finally, a variance analysis and Tukey test were conducted to analyze the results. The results of this work allowed verifying the need to use this additive for the development of the evaluated coatings.

Keywords Composites · Sustainability · Coatings

Introduction

Carbon steel is a material that is used on a large scale in various areas, one of the main ones being in the manufacture of industrial pipes due to their excellent mechanical strength, ductility, malleability, and the fact that it is possible to alter their carbon content, thus optimizing their properties to suit the application [1, 2]. However, these,

D. C. R. Velasco (✉) · D. L. R. Oliveira · F. P. D. Lopes · D. Souza · C. M. F. Vieira
LAMAV—Advanced Materials Laboratory, UENF—State University of the Rio de Janeiro, Av. Alberto Lamego, 2000, Campos dos Goytacazes, Rio de Janeiro 28013-602, Brazil
e-mail: davidc.r.v2014@gmail.com

F. P. D. Lopes
e-mail: perisse@uenf.br

D. Souza
e-mail: djsouza@uenf.br

C. M. F. Vieira
e-mail: vieira@uenf.br

B. Wisner et al. (eds.), *Composite Materials*, The Minerals, Metals & Materials Series,
https://doi.org/10.1007/978-3-031-50180-7_10

especially steel, have a low resistance to corrosive media, which is labeled as the main mechanism that generates failures in pipelines (corrosion), leading to an increase in risk, both economically and in terms of safety [3].

Corrosion in carbon steel pipelines ensues from the influence of diverse corrosive agents, including acids, bases, saline water, and galvanic interactions among dissimilar metals, microbiological activity, atmospheric oxidation, and others. The magnitude of this effect may fluctuate, and it is categorized as either low or high based on standards such as ISO 12944, employing criteria like mass loss or diminishment of the metallic substrate's thickness [4].

One way of trying to mitigate this damage is by applying a coating, which can be used for various purposes, such as to protect parts against corrosive and abrasive agents [5, 6]. These types of coatings have a wide range of applications and are used in industries such as oil and gas, construction, defense, and medical [7–9].

Epoxy-based coatings represent a field of study and development of materials of great interest [10]. This system, often presented in the form of a two-component compound, consists of an epoxy resin and a hardener. When these components are combined, a curing process takes place, forming intermolecular bonds that result in a robust and durable surface. These materials have gained prominence due to their remarkable properties, which make them highly desirable in various industrial applications, such as use in anti-corrosion coatings. Some of these properties include high chemical stability, solid adhesion to metals, and relatively low cost [11, 12].

But this type of epoxy system has an exothermic curing process, which generates heat and gases, thus releasing them in the process [12]. This characteristic can generate a series of problems that are detrimental to a pipe coating, such as the formation of discontinuities such as pores and cracks, which can impair the coating's barrier properties [13].

Various types of anti-corrosion coatings use some material in their dispersed phase, thus being characterized as a composite. That is a combination of two materials of different natures, which aims to optimize performance for a given application [14]. The exacerbated demand for natural resources and concerns about the life cycle of these resources have triggered worries about the sustainability of the current model of society. Especially since this demand almost doubles the amount of resources the planet can supply [15]. To align with this reality, more sustainable processes and materials have been adopted. A notable example of this approach is the inclusion of waste as a component in the formulation of these materials, demonstrating an unequivocal commitment to reducing environmental impact [16].

Some production methods are used in both academia and industry to mitigate problems related to the curing of materials. They are carried out to validate and study the influence of applying new processes at the production stage. One method widely used in industry is the manufacture of composite materials under vacuum [17].

In addition, it is essential to explore alternatives that do not require the use of electricity for their implementation and that, at the same time, are faster. In this context, it is important to consider adopting another method, with the use of anti-bubble additives being an excellent example. During the manufacture of the material, this additive plays a fundamental role in facilitating the removal of bubbles during the curing process and breaking them up when they emerge on the surface of the material [18]. This process is possible due to the additive's ability to reduce the material's surface tension. In addition to reducing surface tension, some additives can release gases, which force the bubbles to break when they penetrate them, as well as changing the viscosity of the material depending on the amount applied. This makes it easier to expel the bubbles during the curing process [19–21], proving to be a viable alternative for industrial application as it does not significantly increase production costs.

The objective of this study is to compare the water vapor permeability of epoxy-based anti-corrosion coatings that have been reinforced with particles of coconut waste and/or chamotte (red ceramic). This comparison will involve the manufacture of coatings both with and without the inclusion of an anti-bubble additive.

Materials and Methods

The epoxy system used in this work is DGEBA/DETA, marketed by AVIPOL and presented as SQ 1005 and SQ 3131. This system was developed for bonding, emergency repair, and component coating, and is suitable for use with reinforcements [22]. The additive used in this work was SILADIT 53, supplied by the manufacturer of the aforementioned resin. It is suitable for epoxy systems and consists of a solution of hydrocarbons and silicone, the aim of which is to reduce the formation of bubbles and foam [23].

To this end, four formulations were evaluated using red ceramic and/or coconut shell waste, as shown in Table 1. This waste was processed in a ball mill for 24 h, using the particulates that passed through a 150 micron opening (100 mesh sieve).

After the initial curing stage, the coatings were subjected to a post-curing process, keeping them at a temperature of 70 °C for a period of 3 h.

The permeability test was conducted using the Payne cup method, in accordance with ASTM D1653 method B [24]. The condition chosen for the test was condition C, the most similar to the environmental conditions, which involved a temperature of approximately 23 °C. As part of the procedure, the three samples per composition were tested simultaneously. The samples were placed in a desiccator and measurements were taken after a period of 24 h, with the total duration of the test extending

Table 1 Values in volume fraction

ID	Epoxy system (%)	Chamotte particulates (%)	Coconut husk particulates (%)
EP	100	–	–
CC	97.5	–	2.5
CH	80	20	–
HB	80	17.5	2.5

over 7 days. The water vapor transmission rate was calculated using Eq. 1.

$$\text{WVT} = \frac{G}{A} \tag{1}$$

where

WVT water vapor transmission rate [g/m^2 24 h];
G angular coefficient of the line with the lowest coefficient of determination as a function of the curve of variation of mass (g) by time (days) [g/24 h];
A test area [mm^2].

The significance of the variation in permeability was assessed using an Analysis of Variance (ANOVA) with a significance level of 5%.

The samples for assessing porosity were made on a metal substrate measuring 200 × 25 × 6 mm and covered with a 0.8 mm coating. Following the application of the coating and its curing, three samples of each formulation were bent using a universal testing machine (INSTRON 5582). The parameters used in the unfolding were: a four-point bending apparatus, with a distance between the cleavers of 40 mm and a feed speed of 40 mm/min, in accordance with procedure B of NACE SP 0394 [25].

In this work, porosity was classified according to NACE SP 0394 and presented through a detailed analysis carried out with an Olympus LEXT confocal microscope, which offered 432× magnification for precise three-dimensional visualization of the sample's characteristics. This combination of methods allowed a thorough assessment of porosity in coatings applied to metallic substrates, following NACE guidelines, and using advanced microscopy technology for an in-depth understanding of the sample's characteristics.

Results and Discussion

The water vapor permeability rates of the samples showed a linear behavior, with an average coefficient of determination (r^2) of 0.97 for the regressions. The average value of the angular coefficient of these regressions, as well as their respective standard deviations, can be seen in Fig. 1. From these results, there was no significant variation in the permeability of the coatings with the use of anti-bubbles, with P values being obtained within a range of 0.23–0.62.

The porosity of the cross section and the interface can be seen in Fig. 2. In the interface section, shown in Fig. 2a, no significant pores were identified in any of the formulations. On the other hand, in the cross section, shown in Fig. 2b, although it is possible to detect the presence of pores, especially in the formulations with the highest number of particulates (CH and HB), this presence is limited and does not seem to decrease. Even if the pores are reduced in size, they can still be seen in

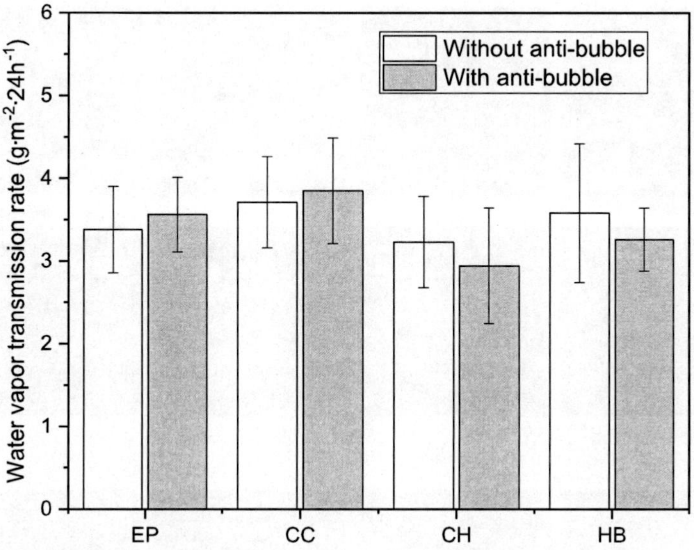

Fig. 1 Permeability of the formulation evaluated

the epoxy system with a similar volume. This explains the lack of variation in the permeability of the coatings that was observed earlier.

About the classification of cross-sectional porosity, it can be concluded that there was no variation in the classification, with all coatings being classified as Grade 1. This means that they have the lowest level of porosity as defined by the NACE SP 0394 standard. This means that, although the additive did not generate significant reductions in porosity, the coating would not fail in this criterion. On the contrary, the degree of roughness is the lowest predicted by NACE.

Conclusions

It can be concluded from this work that the anti-bubble additive had no significant effect on the formulations evaluated, either in terms of porosity or permeability. In the formulations where there was a greater presence of bubbles, the additive acted by breaking them up, but was not effective in eliminating them from the composite.

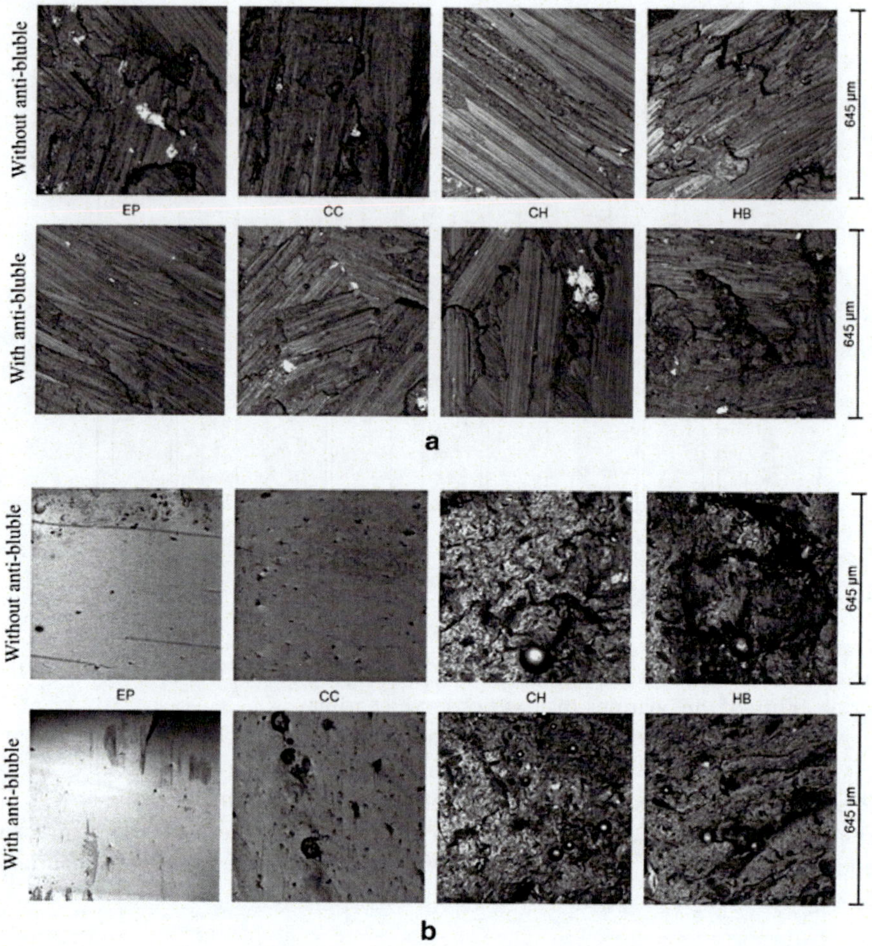

Fig. 2 Porosity of the specimens: **a** interface with substrate; **b** cross section

References

1. Dagdag O, Hsissou R, El HA, Berisha A, Safi Z et al (2020) Fabrication of polymer based epoxy resin as effective anti-corrosive coating for steel: computational modeling reinforced experimental studies. Surf Interfaces. https://doi.org/10.1016/j.surfin.2020.100454
2. Dwivedi D, Lepková K, Becker T (2017) Carbon steel corrosion: a review of key surface properties and characterization methods. RSC Adv 7:4580. https://doi.org/10.1039/C6RA25094G
3. Diraki A, Omanovic S (2022) Smart PANI/epoxy anti-corrosive coating for protection of carbon steel in sea water. Prog Org Coat 168:106835. https://doi.org/10.1016/j.porgcoat.2022.106835
4. International Organization for Standardization (2017) Paints and varnishes—corrosion protection of steel structures by protective paint systems. ISSO 12944. Geneva

5. Croll SG (2020) Surface roughness profile and its effect on coating adhesion and corrosion protection: a review. Prog Org Coat 148:104847. https://doi.org/10.1016/j.porgcoat.2020.105847
6. Zheng S, Yang Y, Wang T, Chen Y, Zhang X et al (2022) High abrasion-resistant biobased composite coatings with improved hardness and corrosion resistance performance based on natural lacquer and SiC nanowires. Prog Org Coat 172:107082. https://doi.org/10.1016/j.porgcoat.2022.107082
7. Faraldos M, Bahamonde A (2019) Multifunctional photocatalytic coatings for construction materials. Nanotechnol Eco-efficient Constr (Second Edition). 2:557–589. https://doi.org/10.1016/B978-0-08-102641-0.00023-2
8. Zhang W, Yu M, Cao Y, Zhuang Z, Zhang K et al (2023) An anti-bacterial porous shape memory self-adaptive stiffened polymer for alveolar bone regeneration after tooth extraction. Bioactive Mater 21:450–463. https://doi.org/10.1016/j.bioactmat.2022.08.030
9. Bobby S, Samad MA (2016) Enhancement of tribological performance of epoxy bulk composites and composite coatings using micro/nano fillers: a review. Polym Adv Technol 28(6):633–644. https://doi.org/10.1002/pat.3961
10. Baig MMA, Samad MA (2021) Epoxy\epoxy composite\epoxy hybrid composite coatings for tribological applications—a review. Polymers 13(2):179. https://doi.org/10.3390/polym13020179
11. Daniyan IA, Mpofu K, Adeodu AO, Adesina O (2021) Development of carbon fibre reinforced polymer matrix composites and optimization of the process parameters for railcar applications. Mater Today Proc 38(2):628–634. https://doi.org/10.1016/j.matpr.2020.03.480
12. Rashid MA, Zhu S, Zhang L, Jin K, Liu W (2023) High-performance and fully recyclable epoxy resins cured by imine-containing hardeners derived from vanillin and syringaldehyde. Eur Polymer J 187:111878. https://doi.org/10.1016/j.eurpolymj.2023.111878
13. Ou B, Wang Y, Lu Y (2020) A review on fundamentals and strategy of epoxy-resin-based anticorrosive coating materials. Polym Plastics Technol Mater 60(6):601–625. https://doi.org/10.1080/25740881.2020.1819317
14. Kerni L, Singh S, Patnaik A, Kumar N (2020) A review on natural fiber reinforced composites. Mater Today Proc 28(3):1616–1621
15. Velasco DCR, Lopes FPD, Souza D, Lopera HAC, Monteiro SN et al (2023) Evaluation of composites reinforced by processed and unprocessed coconut husk powder. Polymers 15(5):1195. https://doi.org/10.3390/polym15051195
16. Bakhori SNM, Hassan MZ, Bakhori NM, Jamaludin KR, Ramlie F et al (2022) Physical, mechanical and perforation resistance of natural-synthetic fiber interply laminate hybrid composites. Polymers 14(7):1322. https://doi.org/10.3390/polym14071322
17. Sarikaya E, Çallioğlu H, Demirel H (2019) Production of epoxy composites reinforced by different natural fibers and their mechanical properties. Compos B 167:461–466. https://doi.org/10.1016/j.compositesb.2019.03.020
18. de Oliveira ML, Reis LCO, Leão RLC, Fragoso MCF, Lima FRA (2020) Solid standards for positron emitters. Radiat Phys Chem 167:108327. https://doi.org/10.1016/j.radphyschem.2019.05.017
19. Jun YD, Kim KJ, Kennedy JM (2010) Dynamic surface tension of heat transfer additives suitable for use in steam condensers and absorbers. Int J Refrig 33(2):428–434. https://doi.org/10.1016/j.ijrefrig.2009.11.006
20. Júnior SVCN (2022) Ciência dos polímeros: um texto básico para tecnólogos e engenheiros. Artiliber poblisher, São Paulo
21. Sritharan R, Askari D (2022) A design of experiment study to investigate the effects of hardener concentration, stirring time, and air bubbles on the tensile strength of epoxy resin. J Elastomers Plast 54(7):1129–1147. https://doi.org/10.1177/00952443221131197
22. SILAEX QUÍMICA (2017) SQ 2050 e SQ 3131. sistema de resina epoxídica. http://www.silaex.com.br/datasheet/2050e3131.pdf. Accessed 2 Sept 2023
23. Folheto de Informações de Segurança de Produtos Químicos (FISPQ): Siladit 53 (2016) SILAEX QUÍMICA. São Paulo

Recovery of Vanadium (IV) from Leaching Solution Using Fe-MOF Material

Wenjuan Wang, Yanfang Huang, and Guihong Han

Abstract Vanadium is an important strategic material that is widely utilized in aerospace, national defense, and metallurgical and chemical engineering fields. V(IV) is a stable form which is derived from the traditional sodium salt roasting-water leaching process for vanadium extraction that widely exists in the leaching solution. To achieve the recovery of V(IV) from the leaching solution, Fe-MOF was employed as an adsorbent for the adsorption of V(IV) from solution in this paper. The influences of reaction time, adsorbent dose, and solution pH on V(IV) adsorption were systematically examined. Under the conditions of initial V(IV) concentrations of 10 mg/L and 100 mg/L, a solution pH of 7, and an adsorbent dosage of 400 mg/L, the removal rates of V(IV) were 97.27% and 52.46%, and the adsorption capacities were 17.62 mg/g and 91.68 mg/g, respectively. The results demonstrated that Fe-MOF can realize the effective recovery of vanadium resources in solution.

Keywords Vanadium · Fe-MOF · Adsorption · Leaching solution

Introduction

Vanadium is an important strategic material that is widely utilized in aerospace, national defense, and metallurgical and chemical engineering fields [1, 2]. Approximately, 88% of vanadium in the world is extracted from vanadium–titanium magnetite ores [3]. In vanadium metallurgy, vanadium–titanium magnetite was melted and oxidatively enriched in a blast furnace to form vanadium slag with a major composition of V(III). Then, the leachable vanadate and/or minor V(IV) were extracted from the vanadium slag by oxidative roasting-leaching methods [4]. The most common oxidation states of vanadium are III, IV, and V in aqueous solution, with the latter two oxidation states being the most stable [5, 6]. However, since highly toxic V(V) always remains in wastewater after vanadate precipitation, it is common

W. Wang · Y. Huang · G. Han (✉)
School of Chemical Engineering, Zhengzhou University, 450001 Zhengzhou, China
e-mail: hanguihong@zzu.edu.cn

B. Wisner et al. (eds.), *Composite Materials*, The Minerals, Metals & Materials Series,
https://doi.org/10.1007/978-3-031-50180-7_11

to detoxify by reducing it to V(IV) before discharge [7]. Therefore, it is important to explore a method for V(IV) recovery from leachate.

Various techniques have been developed to recover ions from solution, such as adsorption, chemical precipitation, ion exchange, and extraction [8, 9]. Among them, adsorption technology is considered the most preferable method due to its high efficiency, low cost, and simple operation conditions. Nevertheless, to the best of our knowledge, few studies have been conducted on the recovery of V(IV) by adsorption. Jansson Charrrier et al. [10] reported the adsorption of vanadium(IV) by chitosan. Vega et al. [11] utilized commercial crystalline calcium hydroxyapatite to remove vanadium(IV) from aqueous solutions. Manohar et al. [9] employed aluminum-pillared clay for the removal of vanadium(IV) from aqueous solution. However, the main obstacles of the above-mentioned adsorbents were poor adsorption and long equilibration times, which greatly limited their application in industry. Therefore, it is important to develop an efficient adsorbent to recover V(IV) from solution.

Recently, more attention has been given to metal–organic frameworks (MOFs) due to their ultra-high porosity, enormous internal surface areas, and tunable pore size. Ahmadijokani et al. [12] synthesized UiO-66-EDA with maximum adsorption capacities of 243.90 mg/g, 217.39 mg/g, and 208.33 mg/g for Pb, Cd, and Cu ions, respectively. Wang et al. [13] prepared water-stable CuII-MOF, which was developed into an effective capture material for the removal of chromate from water. It was found that the adsorption capacity of CuII-MOF was 190 mg/g. Wang et al. [14] reported a promising novel S, N-rich MOF adsorbent for gold recovery. Column adsorption experiments showed that the MOF had outstanding selectivity for Au(III) with 100% removal efficiency in a complex solution containing Au, Ni, Cu, Cd, Co, and Zn ions. These MOF-based porous adsorbents exhibit fast and excellent adsorption abilities for ions removed from solution. However, to date, the employment of Fe-MOF as an adsorbent for the removal of V(IV) from leaching solution has rarely been reported.

In this study, to adequately recover vanadium resources from leaching solution, Fe-MOF was first prepared by a simple solvothermal method, and then the adsorption performance of Fe-MOF on V(IV) was explored. In particular, the effects of different single-factor levels such as adsorbent dose, reaction time, initial concentration of ions, and solution pH on the adsorption behavior of Fe-MOF were mainly examined to provide a basic reference for the recovery of tetravalent vanadium in leaching solution.

Experimental

Materials

Terephthalic acid (H_2BDC, 98%) and iron chloride hexahydrate ($FeCl_3 \cdot 6H_2O$, 98%) were supplied by Shanghai Aladdin Chemistry Co., Ltd. N, N-Dimethylformamide (DMF, 99.5%) and absolute ethanol (>99.9%) were purchased from Tianjin Yongda

Chemical Reagent Co., Ltd. Deionized water was used in all the experiments. Standard V(IV) stock solutions were prepared by dissolving appropriate amounts of vanadyl sulfate pentahydrate ($VOSO_4 \cdot 5H_2O$ in 0.5 mol/L H_2SO_4).

Synthesis of Fe-MOF Adsorbent

The Fe-MOF adsorbent was synthesized according to reported studies [15]. Typically, 3.33 mmol of $FeCl_3 \cdot 6H_2O$ and 1.65 mmol of H_2BDC were dissolved in 60 ml of DMF. The mixture was sonicated for 10 min, and then transferred into a Teflon-lined autoclave and heated at 110 °C for 20 h. After cooling to room temperature, the solid product was obtained by centrifuge, and washed three times with absolute ethanol. Finally, the solid product was dried in a high vacuum at a pressure of over −0.8 bar and a drying temperature of 333.15 K.

Batch Adsorption Experiments

Batch V(IV) adsorption experiments were carried out to investigate the adsorption capacity of the Fe-MOF adsorbent. 200 mg of Fe-MOF adsorbent was placed in a 50 mL conical flask by adding 50 mL of 10 mg/L and 100 mg/L V(IV) solutions. The solution pH was adjusted to 3.0 with 0.1 M HCl and 0.1 M NaOH. Then, the conical flasks were placed in a constant temperature oscillator under constant shaking at a speed of 150 rpm at 25 °C. In this case, the oscillation mode of the thermostatic oscillator used in this study was gyratory oscillation. Afterwards, the mixed solution was filtered using a 0.22 μm polyethersulfone (PES) membrane filter. After filtration, both the initial and residual V(IV) concentrations in the solution were determined by ICP-OES.

The removal efficiency of V(IV) by Fe-MOF was calculated according to Eq. (1).

$$Y = \frac{C_t - C_0}{C_0} \times 100\% \tag{1}$$

where Y (%) is the removal efficiency of V(IV), C_0 (mg/L) is the initial concentration of V(IV), and C_t (mg/L) is the residual concentration of V(IV) at time t.

The adsorption capacity of the Fe-MOF for V(IV) was calculated according to Eq. (2).

$$q_e = \frac{(C_t - C_0)V}{m} \tag{2}$$

where q_e (mg/g) is the adsorption capacity of Fe-MOF for V(IV), V (L) is the solution volume, and m (g) is the mass of the adsorbent.

V(IV) Adsorption Kinetic Model

The mechanism of V(IV) adsorption onto the adsorbents was evaluated by fitting the data into various models. The kinetic data were evaluated by fitting the data onto pseudo-first and pseudo-second kinetic models which are expressed below [16]:

$$\ln(q_e - q_t) = \ln q_e - k_1 t \tag{3}$$

$$\frac{t}{q_t} = \frac{1}{k_2 q_e^2} + \frac{1}{q_e} t \tag{4}$$

where q_t (mg/g) is the adsorption capacity at any time t, and k_1 (1/min) and k_2 (1/min) are the rate constants for the pseudo-first-order and pseudo-second-order kinetic models, respectively.

Results and Discussion

Effect of Reaction Time

To investigate the effect of reaction time on V(IV) removal by the Fe-MOF adsorbent, batch adsorption experiments were carried out in this study at initial V(IV) concentrations of 10 mg/L and 100 mg/L, a solution pH of 3, and an adsorbent dosage of 400 mg/L, and the relevant results are shown in Fig. 1. From Fig. 1, it can be found that with increasing contact time, the removal of V(IV) first increased rapidly and then reached equilibrium. When the initial concentrations of V(IV) were10 mg/L and 100 mg/L and the reaction time was 300 min, the adsorption capacities of Fe-MOF were 13.49 mg/g and 54.65 mg/g, respectively, and the corresponding removal efficiencies were 71.71% and 30.90%, respectively. Compared with conventional adsorbents [9–11], Fe-MOF requires less time for the adsorption equilibrium of V(IV) and has a higher removal rate.

According to Fig. 1, the adsorption of V(IV) on the Fe-MOF adsorbent reached equilibrium after 300 min. Figure 2 presents the experimental data and the simulation results of the pseudo-first-order and second-order kinetic models. The best-fit model parameters are listed in Table 1. From Fig. 2 and Table 1, the R^2 values of the pseudo-second-order model were 0.9981 and 0.9462 when the initial concentrations of V(IV) were 10 mg/L and 100 mg/L, respectively, which better simulated the experimental data. The applicability of the pseudo-second-order model suggested that the adsorption of V(IV) by Fe-MOF may be conducted through a chemisorption mechanism.

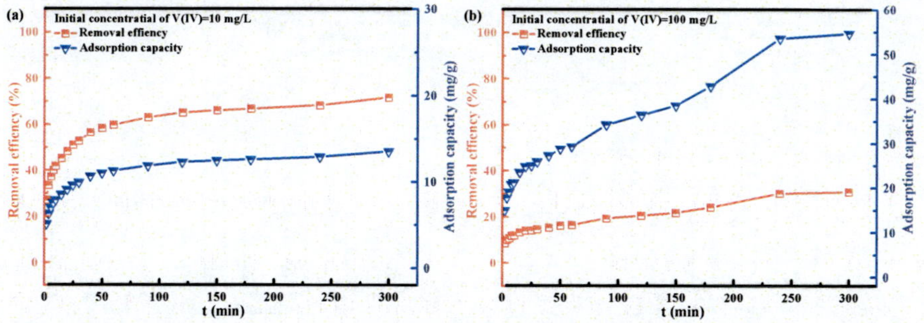

Fig. 1 Effect of contact time on V(IV) adsorption at initial V(IV) concentration of **a** 10 mg/L and **b**100 mg/L (adsorbent dose = 400 mg/L, solution pH = 3)

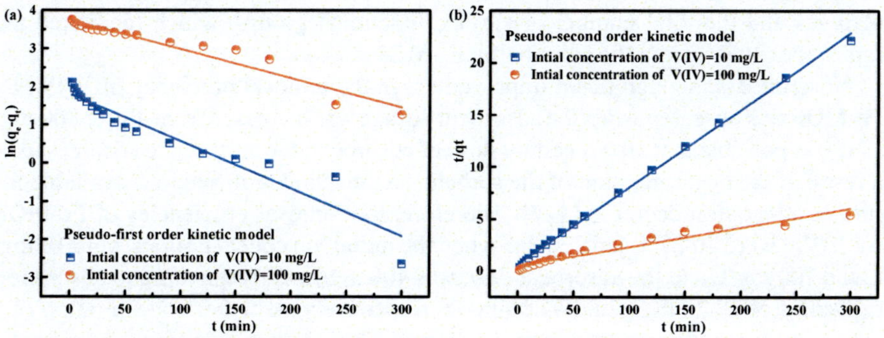

Fig. 2 Adsorption kinetic for V(IV) on Fe-MOF

Table 1 Kinetic parameters of V(IV) adsorption on the Fe-MOF from the adsorption kinetic models

V(IV)	Pseudo-first-order model			Pseudo-second-order model		
	q_e (mg/g)	k_1 (1/min)	R^2	q_e (mg/g)	k_2 (1/min)	R^2
10 mg/L	6.211	0.0124	0.9134	13.46	0.000839	0.9981
100 mg/L	43.055	0.0076	0.9306	53.94	0.0000660	0.9462

Effect of Adsorbent Dose

Adsorbent dose is also an important parameter of the adsorption process, to achieve optimum adsorption and to strike a balance between economic and environmental aspects, the effect of adsorbent dosage was explored and the relevant results are shown in Fig. 3. The results showed that the removal efficiency of Fe-MOF increased with increasing adsorbent dose. This was attributed to the fact that the improvement of Fe-MOF dose increased the corresponding active sites, including the total number

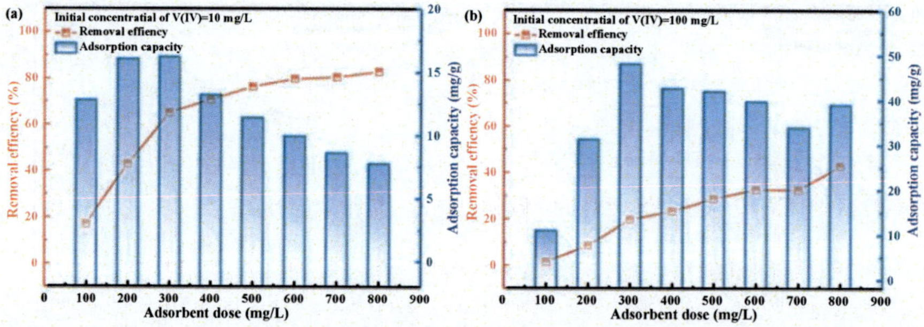

Fig. 3 Effect of absorbent dose of Fe-MOF on V(IV) adsorption at initial V(IV) concentration of **a** 10 mg/L and **b** 100 mg/L (contact time = 120 min, solution pH = 3)

of pores and the total number of surface functional groups, which facilitated the removal of V(IV) from the solution by Fe-MOF.

Nevertheless, no significant improvement in the removal efficiency of V(IV) by Fe-MOF was observed when the adsorbent dosage was increased from 400 to 800 mg/L. It was possible that after a certain dose of adsorbent, the surface area decreased as a result of the agglomeration of the sorbent, so the number of free ions available for further adsorption decreased [17]. Therefore, the removal efficiencies of Fe-MOF for V(IV) were 70.87% and 23.43% when the initial ion concentrations were 10 mg/L and 100 mg/L and the adsorbent dose was 400 mg/L, corresponding to adsorption capacities of 13.20 mg/g and 42.56 mg/g, respectively. A higher concentration of a substance in a solution means there are more molecules available for adsorption. As a result, in the case of a higher concentration (100 mg/L), the Fe-MOF can adsorb more V(IV) ions onto its surface, leading to a higher absorption capacity. Once the Fe-MOF is saturated, it cannot adsorb more V(IV) ions, even though there are more in the solution. Therefore, at higher concentrations, the removal efficiency might be lower because the adsorbent becomes saturated faster, limiting its ability to remove more molecules from the solution.

Effect of Solution pH

The solution pH was one of the main factors that affected the adsorption capacity. The acidity of the medium is an important factor that determines the species of the metal ion present and the surface charge of the adsorbent. The distribution ratio of V(IV) at different concentrations was analyzed using visual Minteq 3.1 open-source software. As shown in Fig. 4, V(IV) ions mainly existed in the form of VO^{2+}, $H_2V_2O_4{}^{2+}$, and $V(OH)_3{}^+$, which were all positively charged. In addition, the distribution ratio of each existing form of V(IV) changed little with increasing initial concentration of the ions.

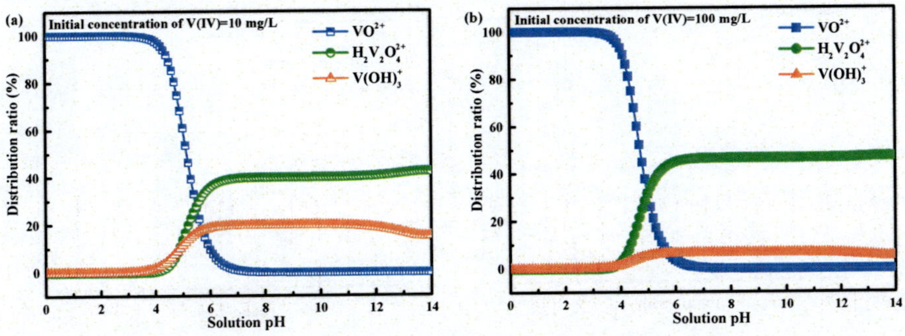

Fig. 4 Distribution ratio diagram of V(IV) at different concentrations

To investigate the effect of solution pH on V(IV) adsorption by the Fe-MOF adsorbent, batch adsorption experiments were carried out at initial V(IV) concentrations of 10 and 100 mg/L, an adsorbent dose of 400 mg/L, and a reaction time of 120 min, and the results are shown in Fig. 5. As shown in Fig. 5, when the solution pH was below 3, the removal efficiency and adsorption capacity of Fe-MOF were very low. This could be attributed to the fact that Fe-MOF undergoes material structural distortion, framework collapse, or even complete degradation when the solution pH is lower than 3 [18]. Structural instability leads to a reduction in the available adsorption surface area and a loss of surface-active adsorption sites, thus reducing the ability of MOF to effectively adsorb pollutants. When the solution pH ranged from 3 to 7, the removal efficiency and adsorption capacity of Fe-MOF increased with increasing solution of pH values. This might be due to the hydrolysis reaction of unsaturated Fe-sites of Fe-MOF [18], which could undergo ligand exchange reactions with V(IV). At the same time, with the increase of pH, the surface of Fe-MOF gradually became negatively charged, resulting in the conversion of electrostatic repulsion to electrostatic attraction between Fe-MOF and V(IV), which was favorable for the recovery of V(IV). When the solution pH was 7 and the initial V(IV) concentrations of 10 and 100 mg/L, the adsorption capacities of Fe-MOF were 17.62 mg/g and 91.68 mg/g, respectively.

Conclusions

In this work, we prepared Fe-MOF by a solvothermal method and successfully utilized it for the recovery of V(IV) from solution. Adsorption parameters such as reaction time, adsorbent dose, and solution pH were investigated. Adsorption kinetic study revealed that the adsorption process followed a pseudo-second-order equation. Batch adsorption experiments indicated that the adsorption process was mainly controlled by chemisorption. Under the conditions of initial V(IV) concentrations of 10 mg/L and 100 mg/L, a solution pH of 7, and an adsorbent dose of 400 mg/L, the

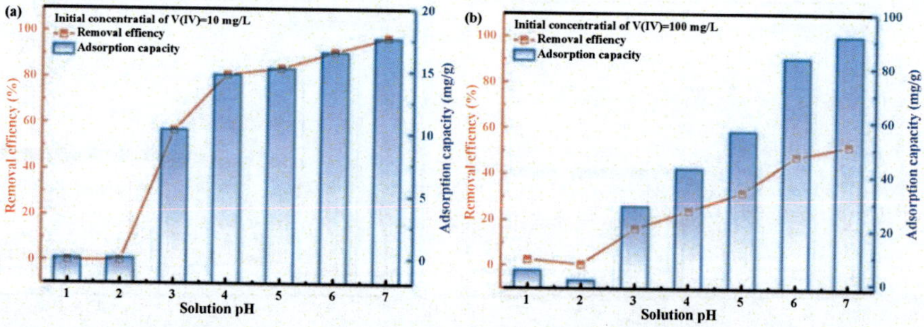

Fig. 5 Effect of solution pH on V(IV) adsorption at initial V(IV) concentration of **a** 10 mg/L and **b** 100 mg/L. (contact time = 120 min, adsorbent dose = 400 mg/L)

removal rates of V(IV) were 97.27% and 52.46%, and the adsorption capacities were 17.62 mg/g and 91.68 mg/g, respectively. The results demonstrated that Fe-MOF can realize the effective recovery of vanadium resources in solution.

Acknowledgements This work was supported by the National Natural Science Foundation of China (NSFC) (No. U2004215, 51974280, U22A20130, 52150079) and Natural Science Foundation of Henan Province of China (No. 232300421196).

References

1. Li H, Wang K, Hua W et al (2016) Selective leaching of vanadium in calcification-roasted vanadium slag by ammonium carbonate. Hydrometallurgy 160:18–25
2. Kang Q, Zhang Y, Bao S (2019) Cleaning method of vanadium precipitation from stripped vanadium solution using oxalic acid. Powder Technol 355:667–674
3. Li H, Li D, Yang Y et al (2020) Re-examination of complexation behaviors of V(V) and V(IV): experimental investigation and theoretical simulation. J Anal At Spectrom 35(5):878–885
4. Wang Z, Chen L, Aldahrib T et al (2020) Direct recovery of low valence vanadium from vanadium slag—Effect of roasting on vanadium leaching. Hydrometallurgy 191:105156
5. Abigail PR, José AHV, José RP et al (2015) Synthesis of protonated chitosan flakes for the removal of vanadium (III, IV and V) oxyanions from aqueous solutions. Microchem J 118:1–11
6. Knežević L, Cukrov N, Bura Nakić E (2020) Ion-exchange chromatography as a tool for investigating vanadium speciation in sediments: preliminary studies. J Soils Sed 20(6):2733–2740
7. Kończyk J, Kluziak K, Kołodyńska D (2022) Adsorption of vanadium (V) ions from the aqueous solutions on different biomass-derived biochars. J Environ Manage 313:114958
8. Bhatnagar A, Minocha AK, Pudasainee D et al (2008) Vanadium removal from water by waste metal sludge and cement immobilization. Chem Eng J 144(2):197–204
9. Manohar DM, Noeline BF, Anirudhan TS (2005) Removal of vanadium(IV) from aqueous solutions by adsorption process with aluminum-pillared bentonite. Ind Eng Chem Res 44(17):6676–6684
10. Jansson Charrier M, Guibal E, Roussy J et al (1996) Vanadium (IV) sorption by chitosan: kinetics and equilibrium. Water Res 30(2):465–475
11. Vega ED, Pedregosa JC, Narda GE et al (2003) Removal of oxovanadium(IV) from aqueous solutions by using commercial crystalline calcium hydroxyapatite. Water Res 37(8):1776–1782

12. Ahmadijokani F, Tajahmadi S, Bahi A et al (2021) Ethylenediamine-functionalized Zr-based MOF for efficient removal of heavy metal ions from water. Chemosphere 264:128466
13. Shao Z, Huang C, Wu Q et al (2019) Ion exchange collaborating coordination substitution: more efficient Cr(VI) removal performance of a water-stable CuII-MOF material. J Hazard Mater 378:120719
14. Wang B, Ma Y, Xu W et al (2023) A novel S, N-rich MOF for efficient recovery of Au(III): performance and mechanism. J Hazard Mater 451:131051
15. Maksimchuk NV, Kovalenko KA, Fedin VP et al (2012) Cyclohexane selective oxidation over metal–organic frameworks of MIL-101 family: superior catalytic activity and selectivity. Chem Commun 48(54):6812–6814
16. da Costa JS, Bertizzolo EG, Bianchini D et al (2021) Adsorption of benzene and toluene from aqueous solution using a composite hydrogel of alginate-grafted with mesoporous silica. J Hazard Mater 418:126405
17. Yousif A, El-Afandy A, Dabbour G et al (2022) Selective separation of V(IV) from its solutions using modified cellulose. J Dispersion Sci Technol 43(10):1427–1437
18. Wang W, Huang Y, Han G et al (2022) Enhanced removal of P(V), Mo(VI) and W(VI) generated oxyanions using Fe-MOF as adsorbent from hydrometallurgical waste liquid: exploring the influence of ionic polymerization. J Hazard Mater 427:128168

Removal of Cr(III) Ions from Simulated Hydrometallurgical Wastewater by Fe-MOF

Hongfei Ma, Lulu Kou, Wenjuan Wang, Yanfang Huang, and Guihong Han

Abstract Wastewater containing Cr(III) ions in hydrometallurgy is likely to cause harm to the surrounding environment and the human body. In this paper, Fe-MOF was synthesized by a solvothermal method and employed as an adsorbent to remove Cr(III) ions from hydrometallurgical wastewater. The effects of initial concentration and pH on the chemical forms of Cr(III) were mainly evaluated, while the influence of adsorption time and solution pH on the removal of Cr(III) ions under different initial concentration conditions were comprehensively analyzed. Under a contact time of 2 h, pH 7.0, and concentration of 100 $mg{\cdot}L^{-1}$, the optimum adsorption efficiency of Cr(III) ions was 82.63%. The adsorption methods of Cr(III) by Fe-MOF were investigated by batch adsorption experiments and kinetic simulations, and the results suggested that Fe-MOF has a promising application in the treatment of high-concentration Cr(III)-containing waste liquids.

Keywords Chromium · Removal of Cr(III) ions · Fe-MOF · Waste liquid

Introduction

Chromium as an important nonferrous metal is extensively used in the metallurgical and chemical industries [1]. Cr(VI) has emerged as the focus of removal studies because of its high toxicity [2]. Although the toxicity of Cr(III) is weaker than that of Cr(VI), Cr(III) in metallurgical wastewater is susceptible to reoxidation to Cr(VI) during the transformation process [3]. In addition, Cr(III) has a teratogenic effect on the human body [4] and destroys the bacteria in the soil, leading to reduced crop yields [5]. Therefore, the danger of trivalent chromium cannot be ignored, and it is also crucial to remove trivalent chromium from metallurgical wastewater. Chromium removal techniques include chemical precipitation [6], electro-flocculation [7], electrodialysis [8], and reverse osmosis [9]. In comparison, the adsorption method is simple in operation, efficient in treatment, and is low cost [10–12].

H. Ma · L. Kou · W. Wang · Y. Huang · G. Han (✉)
School of Chemical Engineering, Zhengzhou University, 450001 Zhengzhou, China
e-mail: hanguihong@zzu.edu.cn

B. Wisner et al. (eds.), *Composite Materials*, The Minerals, Metals & Materials Series,
https://doi.org/10.1007/978-3-031-50180-7_12

Metal–organic frameworks (MOFs) materials are porous metallic materials with high specific surface area, porosity, pore size, and porous structure [13]. In addition, the central metal ions of MOF materials can be regulated. Therefore, MOF materials are considered to be excellent adsorbents, and an increasing number of MOF materials have been investigated for the treatment of heavy metal wastewater [14, 15].

In this work, Fe-MOF was synthesized by a solvothermal method, and the forms of Cr(III) ions in solution were investigated at different pH values and initial concentrations. The batch adsorption experiments and kinetic simulations show that Fe-MOF has good application prospects for the removal of Cr(III) ions.

Experimental

Materials

Terephthalic acid (H_2BDC) and ferric chloride hexahydrate ($FeCl_3 \cdot 6H_2O$) were obtained from Shanghai Aladdin Biochemical Technology Co., Ltd. N, N-dimethylformamide (DMF), anhydrous ethanol (C_2H_5OH), hydrochloric acid (HCl), sodium hydroxide (NaOH), and chromium trichloride hexahydrate ($CrCl_3 \cdot 6H_2O$) were purchased from Tianjin Yunda Chemical Reagent Co., Ltd. All experiments were conducted using analytical grade.

Syntheses of Fe-MOF Adsorbents

The Fe-MOF adsorbent was synthesized based on reported relevant studies [14]. First, 3.33 mmol of $FeCl_3 \cdot 6H_2O$ and 1.65 mmol of H_2BDC were dissolved in 60 mL of DMF. The mixture was sonicated for 30 min before transferring to a Teflon-lined autoclave to be heated at 110 °C for 20 h. The mixture was then cooled to room temperature, centrifuged to separate the solutes, washed three times with anhydrous ethanol, and finally dried under vacuum at 60 °C overnight.

Batch Adsorption Experiments

To study the adsorption capability of Fe-MOF for Cr(III) ions, batch Cr(III) ion adsorption experiments were carried out. Then, 100 mL of 10 $mg \cdot L^{-1}$ and 100 $mg \cdot L^{-1}$ solutions were taken in the conical flasks, respectively, and the pH was adjusted to 1.0, 2.0, 3.0, 4.0, 5.0, 6.0, and 7.0 with HCl and NaOH to explore the effect of pH on adsorption efficiency. Then, 40 $mg \cdot L^{-1}$ Fe-MOF was added to the conical flasks, and the flasks were placed on a thermostatic shaker and shaken at a speed of 150 rpm at 25 °C. The mixed solution was passed through 0.22 μm membrane filters. Finally, the concentrations of Cr(III) ions in the solution before and after adsorption were measured by inductively coupled plasma–optical emission spectrometry (ICP–OES). The removal efficiency (R_t) was calculated by Eq. (1).

$$R_t = \frac{C_0 - C_t}{C_0} \times 100\% \quad (1)$$

where R_t (%) is the removal efficiency at time t, t (min) is the contact time, C_0 (mg·L^{-1}) is the initial Cr(III) concentration, and C_t (mg·L^{-1}) is the Cr(III) concentration at time t.

Results and Discussion

Influence of Different Concentrations and Solution pH Values on the Adsorption Efficiency of Cr(III) Ions

The forms of trivalent chromium ions present in a solution will change when the pH of the solution and the initial concentration of trivalent chromium ions change. Hence, the solution pH is an essential parameter in the effectiveness of adsorption [16] and the study of the forms of Cr(III) ions at different pH values will help us to remove the Cr(III) ions more conveniently. In this paper, the forms of Cr(III) ions at different pH values and initial concentrations were calculated by using Visual MINTEQ software [17] developed under the leadership of Jon Petter Gustafsson. The results are shown in Fig. 1.

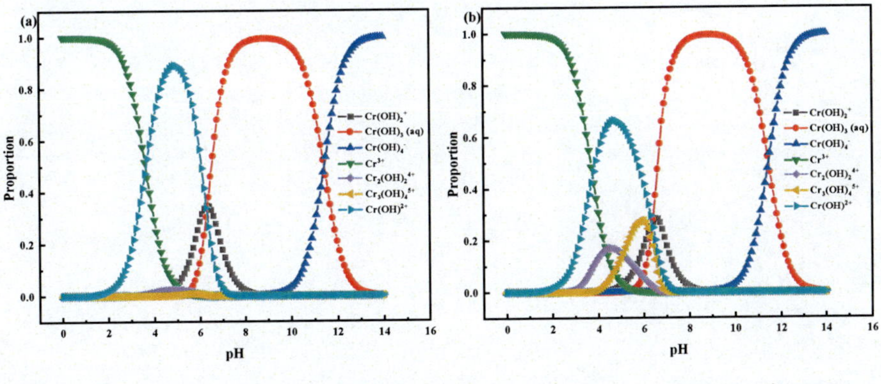

Fig. 1 **a** The forms of 10 mg·L^{-1} Cr(III) ions in solution, **b** The forms of 100 mg·L^{-1} Cr(III) ions in solution

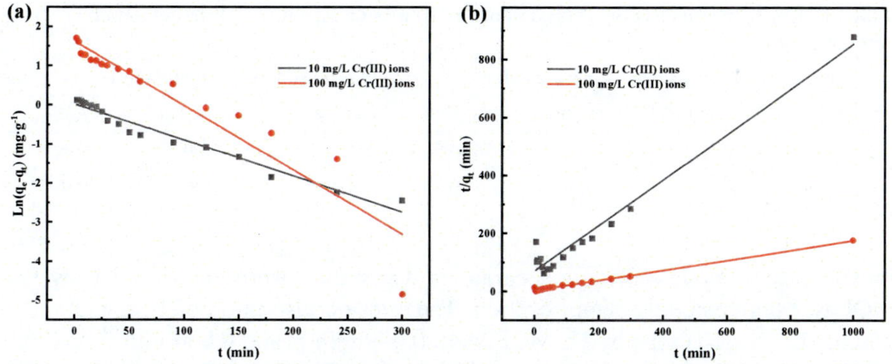

Fig. 3 **a** Quasi-primary kinetic model for the adsorption of Cr(III) ions by Fe-MOF at different initial concentrations, **b** Quasi-secondary kinetic model for the adsorption of Cr(III) ions by Fe-MOF at different concentrations

Kinetic Analysis of the Adsorption of Cr(III) Ions on Fe-MOF

To explore the adsorption mechanism of Cr(III) ions adsorbed on Fe-MOF, quasi-primary and quasi-secondary kinetic models were adopted to fit the experimentally obtained data. The quasi-primary kinetic model indicates that the adsorption of Cr(III) by Fe-MOF is physisorption, while the quasi-secondary kinetic model indicates that adsorption is controlled by chemisorption [19]. The simplified linear quasi-primary kinetic model and quasi-secondary kinetic model are given by Eqs. (2) and (3), and the fitted curves are shown in Fig. 3.

$$\ln q_e - q_t = \ln q_e - K_1 t \tag{2}$$

$$\frac{t}{q_t} = \frac{1}{K_2 q_e^2} + \frac{t}{q_e} \tag{3}$$

where t (min) is the time of adsorption, q_e (mg·g^{-1}) is the adsorbed amount when adsorption reaches equilibrium, q_t (mg·g^{-1}) is the adsorbed amount at the moment t, K_1 (min^{-1}) is the quasi-primary kinetic adsorption rate constant, and K_2 (min^{-1}) is the quasi-secondary kinetic adsorption rate constant.

As indicated in Table 1, the correlation coefficients R^2 of the quasi-secondary kinetic model for the adsorption of Cr(III) ions by Fe-MOF were 0.966 and 0.995 at different initial concentrations which were higher than the correlation coefficients 0.960 and 0.884 of the quasi-primary kinetic model for the adsorption of Cr(III) ions. The equilibrium adsorption amount when the quasi-secondary kinetic model reaches adsorption equilibrium was also more closely related to the actual measurement than the quasi-primary kinetic model. It could be recognized that the quasi-secondary kinetic model for the adsorption of Cr(III) ions by Fe-MOF at different

Table. 1 Kinetic parameters of adsorption of Cr(III) ions on Fe-MOF at different initial concentrations

C_0 (Cr(III)) ($mg \cdot L^{-1}$)	Quasi-primary kinetic model			Quasi-secondary kinetic model		
	q_e ($mg \cdot g^{-1}$)	K_1 (min^{-1})	R^2	q_e ($mg \cdot g^{-1}$)	K_2 (min^{-1})	R^2
10	1.022	0.009	0.960	1.274	0.009	0.966
100	5.087	0.016	0.884	5.818	0.006	0.995

initial concentrations was more accurate, in which the adsorption of Cr(III) by Fe-MOF was dominated by chemisorption. The rates of adsorption of 10 $mg \cdot L^{-1}$ and 100 $mg \cdot L^{-1}$ Cr(III) ions by Fe-MOF were 0.009 min^{-1} and 0.006 min^{-1}, respectively. It could be concluded that the adsorption of Cr(III) in solution by Fe-MOF was attributed to the combination of chemisorption and electrostatic action.

Conclusions

In this work, Fe-MOF was fabricated by the solvothermal method and successfully applied for the removal of Cr(III) ions from the solution. Based on the batch adsorption experiments and kinetic simulations, it could be concluded that the adsorption of Cr(III) ions by Fe-MOF was based on a combination of electrostatic attraction and chemisorption. Due to the combined effect of electrostatic attraction and chemisorption, the adsorption efficiency of Cr(III) ions with the Fe-MOF adsorbent was 82.63% when the adsorbent dose was 400 $mg \cdot L^{-1}$, the initial Cr(III) ion concentration was 100 $mg \cdot L^{-1}$, pH = 7.0, and the contact time was 120 min. The results demonstrated that Fe-MOF had broad application prospects in the treatment of high-concentration Cr(III) ion-containing waste liquids.

Acknowledgements This work was supported by the National Natural Science Foundation of China (NSFC) (No. U2004215, 51974280, U22A20130, 52150079) and Natural Science Foundation of Henan Province of China (No. 232300421196).

References

1. Tang XY, Wu HF, Quan X et al (2020) Separation of sodium chromate from high alkaline media based on precipitation transformation using barium hydroxide. New J Chem 44(28):12048–12057
2. Feng YX, Yu XZ, Mo CH et al (2019) Regulation network of sucrose metabolism in response to trivalent and hexavalent chromium in oryza sativa. J Agric Food Chem 67(35):9738–9748
3. Zhang W, Ma XY, Li R et al (2021) Rapid sequestration of chelated Cr(III) by ferrihydrite: adsorption and overall transformation of Cr(III) complexes. Colloid Surface A 625:126819

4. Ratnalu GV, Dhakate R (2021) Human health hazard evaluation with reference to chromium (Cr^{+3} and Cr^{+6}) in groundwater of Bengaluru Metropolitan City, South India. Arabian J Geosci 14(23):2472
5. Fatima HE, Ahmed A (2018) Micro-remediation of chromium contaminated soils. PeerJ 6:e6076
6. Jiang B, Niu Q, Li C et al (2020) Outstanding performance of electro-Fenton process for efficient decontamination of Cr(III) complexes via alkaline precipitation with no accumulation of Cr(VI): important roles of iron species. Appl Catal B 272:119002
7. Li JH, Bai J, Huang K et al (2014) Removal of trivalent chromium in the complex state of trivalent chromium passivation wastewater. Chem Eng J 236:59–65
8. Ohira S, Nakamura K, Shelor CP et al (2015) Simultaneous electrodialytic preconcentration and speciation of chromium(III) and chromium(VI). Anal Chem 87(22):11575–11580
9. Engstler R, Reipert J, Karimi S et al (2022) A Reverse osmosis process to recover and recycle trivalent chromium from electroplating wastewater. Membranes 12(9):853
10. Adil HI, Thalji MR, Yasin SA et al (2022) Metal–organic frameworks (MOFs) based nanofiber architectures for the removal of heavy metal ions. RSC Adv 12(3):1433–1450
11. Rezaei H (2016) Biosorption of chromium by using Spirulina sp. Arab J Chem 9(6):846–853
12. Teshale F, Karthikeyan R, Sahu O (2020) Synthesized bioadsorbent from fish scale for chromium (III) removal. Micron 130:102817
13. Li ZC, Liu XM, Jin W et al (2019) Adsorption behavior of arsenicals on MIL-101(Fe): The role of arsenic chemical structures. J Colloid Interface Sci 554:692–704
14. Maksimchuk NV, Kovalenko KA, Fedin VP et al (2012) Cyclohexane selective oxidation over metal-organic frameworks of MIL-101 family: superior catalytic activity and selectivity. ChemComm 48(54):6812–6814
15. Jia DM, Li YJ, Cai HM et al (2023) MIL-101(Fe) metal–Organic framework nanoparticles functionalized with amino groups for Cr(VI) capture. ACS Appl Nano Mater 6(8):6820–6830
16. Kanagaraj J, Senthilvelan T, Panda RC et al (2014) Biosorption of trivalent chromium from wastewater: an approach towards green chemistry. Chem Eng Technol 37(10):1741–1750
17. Yi Y, Wang XY, Ma J et al (2014) An efficient egeria najas-derived biochar supported nZVI composite for Cr(VI) removal: characterization and mechanism investigation based on visual MINTEQ model. Environ Res 189:109912
18. Wang WJ, Huang YF, Han GH et al (2022) Enhanced removal of P(V), Mo(VI) and W(VI) generated oxyanions using Fe-MOF as adsorbent from hydrometallurgical waste liquid: exploring the influence of ionic polymerization. J Hazard Mater 427:128168
19. Yu J, Hao HQ, Fan GX et al (2022) Combined experimental and molecular simulation investigation of dodecylamine adsorption onto MIL-100(Fe) for wastewater treatment. J Environ Chem Eng 10(6):108756

Study on the Adsorption of Selenium-Containing Wastewater by MIL-101-NH_2

Lulu Kou, Hongfei Ma, Wenjuan Wang, Yanfang Huang, and Guihong Han

Abstract Large quantities of selenium (Se)-containing wastewater are generated by the coal mining process. Therefore, it is urgent to treat wastewater containing high concentrations of Se(IV). In this study, MIL-101-NH_2 was synthesized by a hydrothermal method, and the effects of parameters, such as pH and contact time on the removal of Se(IV), were evaluated. The adsorption kinetics of MIL-101-NH_2 on Se(IV) were evaluated using pseudo-first-order and pseudo-second-order models. Langmuir and Freundlich models were used to analyze the equilibrium isotherms of the system. The maximum adsorption was 32.289 mg/g according to the Langmuir model. The results showed that the removal efficiency of Se(IV) by MIL-101-NH_2 was 75.96% at pH = 6 and adsorption time = 3 h. This study demonstrated that MIL-101-NH_2 has great potential for wastewater treatment.

Keywords MIL-101-NH_2 · Se(IV) · Adsorption · Wastewater treatment

Introduction

Selenium (Se) is a nutrient required by humans and animals in small amounts. However, high concentrations of Se are highly toxic and can lead to selenium poisoning or death [1]. Excessive human intake of Se (400 μg/day) increases the risk of neurological disorders, including peripheral nerve changes, hair loss, and skin damage [2]. Se is a trace element associated with sulfur-containing minerals in nature. Most of the Se in water is a byproduct of human activities, including mining, oil refining, pesticide production, and other processes [3]. Se(IV) has higher toxicity than Se(VI) in oxidized wastewater and is present in oxidized wastewater in the form of selenite [4].

Commonly used methods for removing Se(VI) from wastewater include chemical reduction [2], bioremediation [1, 5], artificial wetland [4], and adsorption [6]. Among these methods, adsorption is considered an important technology for Se(VI) removal

L. Kou · H. Ma · W. Wang · Y. Huang · G. Han (✉)
School of Chemical Engineering, Zhengzhou University, 450001 Zhengzhou, China
e-mail: hanguihong@zzu.edu.cn

B. Wisner et al. (eds.), *Composite Materials*, The Minerals, Metals & Materials Series,
https://doi.org/10.1007/978-3-031-50180-7_13

because of its high efficiency, economy, and simplicity of operation. In addition, some of the adsorbents can be recycled, which makes the adsorption method most likely to be sustainable. Currently, commonly used adsorbents, such as activated carbon [7, 8], manganese-iron oxides [6], modified zeolites [9], iron oxides [10], and aluminum oxides [11], are still limited in practical applications due to their poor adsorption capacity and long equilibration times. Ma et al. [12] used BUC-17 to remove ReO_4^- and found that the maximum adsorption capacity of BUC-17 was about 409.1 mg/g and the adsorption equilibrium time was 100 min. Wang et al. [13] used MIL-101 to remove Mo(VI) oxyanions and found that the maximum adsorption capacity of Fe-MOF was about 352 mg/g and the adsorption equilibrium time was 30 min. Hence, it is crucial to develop new adsorbents with high adsorption efficiency, easy regeneration, and simple recovery.

Compared with traditional adsorbents, metal–organic frameworks (MOFs) have the advantages of a simple synthesis method, large specific surface area, high removal efficiency, strong adsorption, and regeneration ability [14]. MIL-101-NH_2 is one of the typical Fe-MOF materials with excellent porosity and high specific surface area, which is commonly used in wastewater treatment [15, 16]. In addition, according to research, there is a high affinity between Fe and Se [6, 10]. Therefore, MIL-101-NH_2 has great potential for the removal of Se(VI).

In this study, MIL-101-NH_2 was synthesized by a hydrothermal method and used as an adsorbent to remove Se(VI) from simulated wastewater. The effects of solution pH and contact time were also discussed. Furthermore, the effect of MIL-101-NH_2 on the removal of Se(VI) was analyzed by using kinetics and isotherms.

Experimental

Materials

All reagents used in this study were of analytical grade. Ferric chloride hexahydrate ($FeCl_3 \cdot 6H_2O$) was supplied by Kermel Chemical Reagent Co., Ltd. (Tianjin, China). Selenium standard solution, 2-aminoterephthalic acid ($C_8H_7NO_4$), hydrochloric acid (HCl), and sodium hydroxide (NaOH) were supplied by Aladdin Biochemical Technology Co., Ltd. (Shanghai, China). N,N-dimethylformamide (C_3H_7NO) and absolute ethanol (CH_3CH_2OH) were purchased from Yongda Chemical Reagent Co., Ltd. (Tianjin, China).

Synthesis of MIL-101-NH_2

MIL-101-NH_2 materials were prepared by solvothermal synthesis according to previously reported methods [17]. Firstly, 0.906 g (5.0 mmol) of 2-aminoterephthalic acid,

2.702 g (10.0 mmol) of $FeCl_3 \cdot 6H_2O$, and 60 mL of N,N-dimethylformamide were thoroughly mixed and transferred to a reactor heated at 110 °C for 20 h. The solid was purified by centrifugation and then washed three times with absolute ethanol. Finally, reddish-brown powder was obtained after drying at 60 °C for 12 h.

Adsorption Experiments

Batch adsorption experiments were carried out in 100 mL conical flasks. MIL-101-NH_2 was added to an initial Se(VI) concentration of 10 mg/L at a dosage of 200 mg/L adsorbent. Before adding the adsorbent, the solution pH was adjusted with 1 M NaOH and HCl. Then, the mixed solution was reacted on a shaker at 25 °C at 150 rpm. The mixed solution in the shaker oscillates back and forth in a horizontal plane with no amplitude. At the end of the adsorption process, the mixed solution was purified by filtration using a 0.22 μm filter membrane. The filter membrane consisted of polyethersulfone material, which had excellent resistance to acids, alkalis, and other inorganic solvents and would not affect the experimental results. The residual Se(VI) concentration was determined by ICP–OES. The removal efficiency (R_t) was calculated by Eq. (1).

$$R_t = \frac{C_0 - C_t}{C_0} \times 100\% \quad (1)$$

where R_t (%) is the removal efficiency at time t, t (min) is the contact time, C_0 (mg/L) is the initial Se(VI) concentration, and C_t (mg/L) is the Se(VI) concentration at time t.

The adsorption capacity (q_e) was calculated according to Eq. (2).

$$q_e = \frac{V(C_0 - C_t)}{m} \quad (2)$$

where q_e (mg/g) is the adsorption capacity at the equilibrium time, V(L) is the solution volume, and m (g) is the mass of the adsorbents.

Results and Discussion

Effect of Solution pH

Solution pH is an important parameter affecting adsorption [13]. In this study, simulated wastewater was prepared by diluting a standard solution of Se, and MIL-101-NH_2 was used as an adsorbent to remove Se(VI) from simulated wastewater. The removal efficiency of Se(VI) was investigated in the pH range of 2–10 and the results

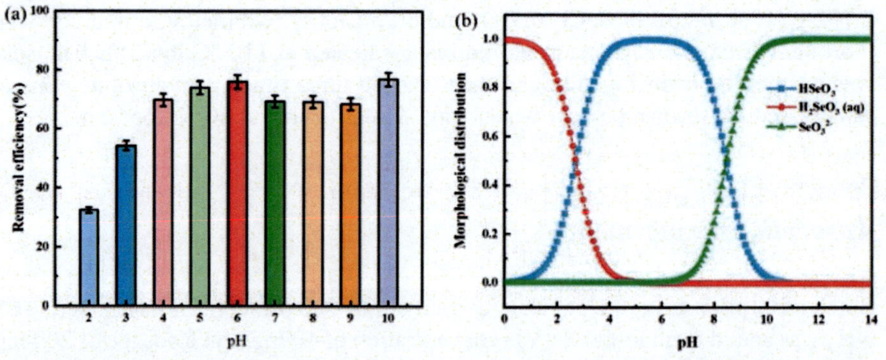

Fig. 1 **a** The effect of solution pH on the removal of Se(IV) (C_0 = 10 mg/L; adsorbent dosage = 200 mg/L; and adsorption time = 3 h); **b** morphological distribution of Se(IV) under different pH conditions

are shown in Fig. 1a. It is clear that pH plays a key role in the adsorption of Se(VI) by MIL-101-NH_2. The highest selenium removal of 75.96% was observed at pH = 6 from Fig. 1a. Therefore, the optimal solution pH was 6.

The ionic form of Se(IV) at a concentration of 10 mg/L simulated using Visual MINTEQ software was displayed in Fig. 1b The solution consisted of H_2SeO_3(ag), $HSeO_3^-$, and SeO_3^{2-}. The lower removal efficiency for solutions with pH values between 2 and 4 could be attributed to the fact that the solutions are only partially negatively charged. When the pH is in the range of 2–6, the solution pH was less than pH_{PZC}, the MIL-101-NH_2 particles were positively charged and attracted the negatively charged Se(VI). When the pH is in the range of 7–9, the solution pH was greater than pH_{PZC}, the MIL-101-NH_2 particles were negatively charged and repelled the negatively charged Se(VI). As the pH = 10, MIL-101-NH_2 may be decomposed, at this time Fe^{3+} adsorption of Se(VI) in the solution leads to a higher removal efficiency.

Effect of Contact Time

Contact time is one of the most important factors affecting the removal efficiency. To understand the effect of contact time, the removal of Se(IV) from simulated wastewater in the range of 0–3 h was studied at an initial Se(IV) concentration of 10 mg/L, the solution pH of 6, and the dosage of 200 m/L of MIL-101-NH_2. From Fig. 2, the removal of Se(IV) increased with time, and equilibrium was essentially reached after 3 h, with a slow increase in the removal efficiency.

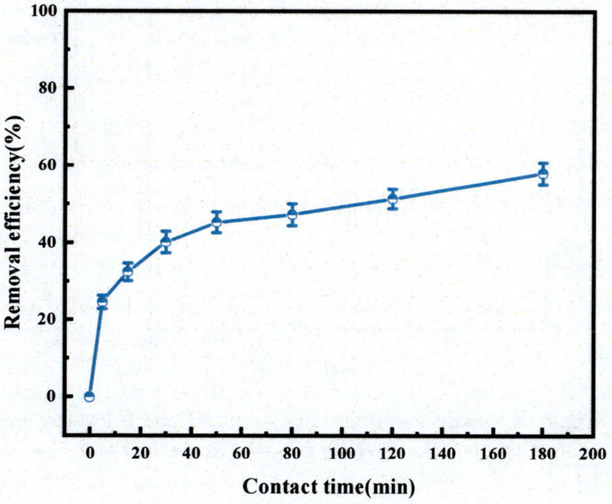

Fig. 2 The effect of contact time on the removal of Se(IV) ($C_0 = 10$ mg/L, adsorbent dosage = 200 mg/L, and solution pH = 6)

Adsorption Kinetics

The adsorption rate is an important factor to evaluate the adsorption performance of an adsorbent. The removal of selenium by MIL-101-NH_2 was investigated by kinetic experiments. Adsorption kinetic modeling was usually performed using the proposed first- and second-order models [18–20]. To analyze the adsorption kinetics, the adsorption kinetic data were processed using the proposed first- and second-order models These fitting equations were given by Eqs. (3) and (4).

Pseudo-first order:

$$\ln{(q_e - q_t)} = \ln q_e - \mathrm{k}_1 t \tag{3}$$

Pseudo-second order:

$$\frac{t}{q_t} = \frac{1}{\mathrm{k}_2 q_e^2} + \frac{t}{q_e} \tag{4}$$

where t (min) is the contact time, qt (mg/g) is the adsorption capacity of the adsorbent at t, and k_1 (1/min) and k_2 (g/mg/min) are the adsorption rate constants for quasi-primary and quasi-secondary kinetics, respectively.

It can be seen from Fig. 3 and Table 1 that the correlation coefficient was 0.794 for the quasi-first-order kinetic fit and 0.991 for the quasi-second-order kinetic fit. It was clear that the pseudo-second-order kinetic model better describes this adsorption process. The qe fit of the pseudo-second-order model was close to the experimental values with the highest correlation coefficient R^2. The applicability of the pseudo-second-order kinetic model indicated that the adsorption of Se(VI) by MIL-101-NH_2 may be conducted through a surface chemisorption mechanism [15].

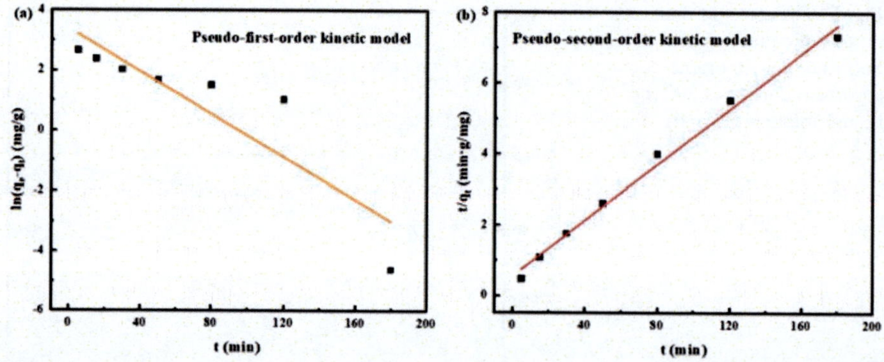

Fig. 3 **a** Pseudo-first-order kinetic model and **b** Pseudo-second-order kinetic model for Se(IV) adsorption on MIL-101-NH_2 ($C_0 = 10$ mg/L, pH = 6, $V = 100$ mL, adsorbent dosage = 200 mg/L)

Table 1 Pseudo-first-order and pseudo-second-order model parameters for Se(VI) adsorption on MIL-101-NH_2

Adsorbent	$q_{e(\text{exp})}$ (mg/g)	Pseudo-first-order kinetic model			Pseudo-second-order kinetic model		
		K_1 (1/min)	q_e (mg/g)	R^2	K_2 (g/mg/min)	q_e (mg/g)	R^2
MIL-101-NH_2	24.560	0.035	29.517	0.794	0.003	25.445	0.991

Adsorption Isotherms

The maximum adsorption capacity is a measure of a good adsorbent. The adsorption capacity of MIL-101-NH_2 for Se(VI) was evaluated by isothermal adsorption experiments. Langmuir and Freundlich models were commonly used to fit adsorption isotherms. The Langmuir model was an ideal adsorption model that assumed adsorption as monomolecular layer adsorption. The Freundlich model was an empirical model of a non-ideal system. To obtain the maximum adsorption capacity, the proposed Langmuir and Freundlich models were used to deal with the data [6, 7]. The Langmuir and Freundlich models were given by Eqs. (5). and (6).

Langmuir model:

$$\frac{C_e}{q_e} = \frac{1}{bq_m} + \frac{C_e}{q_m} \tag{5}$$

Freundlich model:

$$\ln q_e = \ln K_f + \frac{1}{n\ln C_e} \tag{6}$$

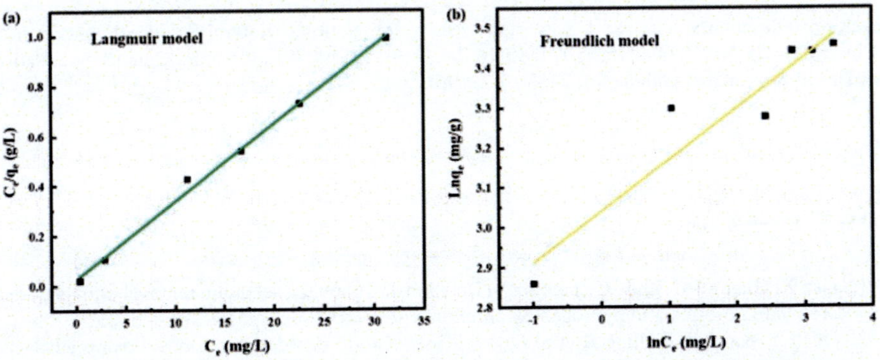

Fig. 4 **a** Langmuir model and **b** Freundlich model for Se(IV) adsorption on MIL-101-NH_2 (C_0 = 5–50 mg/L, pH = 6, V = 50 mL, adsorbent dosage = 200 mg/L, and adsorption time = 3 h)

Table 2 Langmuir and Freundlich isotherm parameters for Se(IV) adsorption on MIL-101-NH_2

Adsorbent	Langmuir model			Freundlich model		
	b (L/mg)	q_m (mg/g)	R^2	K_f [(mg/g)(L/mg)$^{1/n}$]	n	R^2
MIL-101-NH_2	1.133	32.289	0.995	20.916	7.766	0.901

where C_e (mg/L) is the Se(VI) concentration at the equilibrium time, b (L/mg) is the Langmuir constant, K_f [(mg/g)(L/mg)1/n] is the Freundlich constant, and n is a parameter related to the adsorption strength.

According to Fig. 4 and Table 2, among the two isotherm models, the Langmuir model better describes the adsorption process of MIL-101-NH_2 on Se(VI). The correlation coefficient for the Langmuir model fit was 0.995 and the Freundlich model fit was 0.901. The Langmuir model better describes this adsorption process. This indicated that the adsorption process of MIL-101-NH_2 on Se(VI) was monolayer adsorption [21], and the maximum adsorption amount was 32.289 mg/g.

Conclusions

In this work, MIL-101-NH_2 was synthesized by a hydrothermal method and used for the removal of Se(IV) from simulated wastewater. The removal of Se(IV) by MIL-101-NH_2 reached 75.96% at pH = 6, adsorbent dosage = 200 mg/L, and adsorption time = 3 h. The adsorption kinetics and adsorption isotherms showed that the adsorption of Se(IV) by MIL-101-NH_2 was monolayer chemisorption with a maximum adsorption of 32.289 mg/g. This suggests that MIL-101-NH_2 is one of the most promising adsorbent materials with great potential for application in the removal of Se(IV) from wastewater.

Acknowledgements This work was supported by the National Natural Science Foundation of China (NSFC) (No. U2004215, 51974280, U22A20130, 52150079) and Natural Science Foundation of Henan Province of China (No. 232300421196).

References

1. Gan X, Huang J-C, Zhou C et al (2019) Relationship between selenium removal efficiency and production of lipid and hydrogen by chlorella vulgaris. Chemosphere 217:825–832
2. Duan Y, Kaushik V, Jung B et al (2017) Kinetic study of selenium removal using advanced reduction process with dithionite. Environ Sci Technol 35(3):169–175
3. Lu Z, Yu J, Zeng H et al (2017) Polyamine-modified magnetic graphene oxide nanocomposite for enhanced selenium removal. Sep Purif Technol 183:249–257
4. Zhao Q, Huang J-C, He S et al (2020) Enhancement of a constructed wetland water treatment system for selenium removal. Sci Total Environ 714:136741
5. Liu F, Huang J-C, Zhou C et al (2019) Development of an algal treatment system for selenium removal: effects of environmental factors and post-treatment processing of Se-laden algae. J Hazard Mater 365:546–554
6. Qureshi SS, Memon SA, Rafi Ul Z et al (2022) Rapid adsorption of selenium removal using iron manganese-based micro adsorbent. Sci Rep 12(1):17207
7. Okonji SO, Dominic JA, Pernitsky D et al (2020) Removal and recovery of selenium species from wastewater: adsorption kinetics and co-precipitation mechanisms. J. Water Process Eng. 38:101666
8. Kwon JH, Wilson LD, Sammynaiken R (2015) Sorptive uptake of selenium with magnetite and its supported materials onto activated carbon. J Colloid Interface Sci 457:388–397
9. Zhang X, Li X, Jin Z et al (2021) Selenium in wastewater can be adsorbed by modified natural zeolite and reused in vegetable growth. Sci Prog 104(2):00368504211019845
10. Su C, Suarez DL (2000) Selenate and selenite sorption on iron oxides. Soil Sci Soc Am J 64(1):101–111
11. Jegadeesan G, Mondal K, Lalvani SB (2003) Comparative study of selenite adsorption on carbon based adsorbents and activated alumina. Environ Technol 24(8):1049–1059
12. Ma J, Wang C-C, Zhao Z-X et al (2021) Adsorptive capture of perrhenate (ReO_4^-) from simulated wastewater by cationic 2D-MOF BUC-17. Polyhedron 202:115218
13. Wang W, Huang Y, Han G et al (2022) Enhanced removal of P(V), Mo(VI) and W(VI) generated oxyanions using Fe-MOF as adsorbent from hydrometallurgical waste liquid: exploring the influence of ionic polymerization. J Hazard Mater 427:128168
14. Miao Q, Jiang L, Yang J et al (2022) MOF/hydrogel composite-based adsorbents for water treatment: a review. J. Water Process Eng. 50:103348
15. Liu M, Huang Q, Li L et al (2022) Cerium-doped MIL-101-NH_2(Fe) as superior adsorbent for simultaneous capture of phosphate and As(V) from Yangzonghai coastal spring water. J Hazard Mater 423:126981
16. Chen J, Qin C, Mou Y et al (2023) Linker regulation of iron-based MOFs for highly effective Fenton-like degradation of refractory organic contaminants. Chem Eng J 459:141588
17. Yan D, Gao N, Wu X et al (2021) Fabrication of hydrophobic Fe-based metal-organic framework through post-synthetic modification: improvement of aqueous stability. J. Water Process Eng. 40:101979
18. Moore RC, Rigali MJ, Brady P (2016) Selenite sorption by carbonate substituted apatite. Environ Pollut 218:1102–1107
19. Wasewar KL, Prasad B, Gulipalli S (2009) Adsorption of selenium using bagasse fly ash. Clean 37(7):534–543

20. Rovira M, Gimenez J, Martinez M et al (2008) Sorption of selenium(IV) and selenium(VI) onto natural iron oxides: goethite and hematite. J Hazard Mater 150(2):279–284
21. Li D, Yan W, Guo X et al (2020) Removal of selenium from caustic solution by adsorption with CaAl layered double hydroxides. Hydrometallurgy 191:105231

Author Index

B. Wisner et al. (eds.), *Composite Materials*, The Minerals, Metals & Materials Series,
https://doi.org/10.1007/978-3-031-50180-7

Subject Index

B. Wisner et al. (eds.), *Composite Materials*, The Minerals, Metals & Materials Series,
https://doi.org/10.1007/978-3-031-50180-7